Building Inclusive Scientific Communities and Leadership

Building Inclusive Scientific Communities and Leadership

Case Studies from Professional Societies

Edited by
Verónica A. Segarra, Mercy Mugo, and Simone B. Soso

BLOOMSBURY ACADEMIC
NEW YORK • LONDON • OXFORD • NEW DELHI • SYDNEY

BLOOMSBURY ACADEMIC
Bloomsbury Publishing Inc, 1359 Broadway, New York, NY 10018, USA
Bloomsbury Publishing Plc, 50 Bedford Square, London, WC1B 3DP, UK
Bloomsbury Publishing Ireland, 29 Earlsfort Terrace, Dublin 2, D02 AY28, Ireland

First published in the United States of America 2026

Library of Congress Control Number: 2025045229

ISBN: HB: 978-1-6669-7583-3
ePDF: 979-8-2162-5222-1
eBook: 978-1-9787-6697-6

Typeset by Integra Software Services Pvt. Ltd.

Contents

Figures

Tables

Editors and Contributors

Editors

Mercy Mugo (PhD) is Executive Director of the Quality Education for Minorities Network. She has expertise in successfully leading the implementation of capacity building and educational research projects centered in advancing the participation of groups historically underrepresented in STEM education and careers. She leads several initiatives that engage key stakeholders (administrators, faculty, and students), primarily at Minority Serving Institutions (MSIs), to contribute to and benefit from STEM. She has coauthored several articles focused on understanding the learning environments that successfully promote STEM education and research at historically Black colleges and universities.

Verónica A. Segarra (PhD) is Associate Professor and the Maryland E-Nnovation Endowed Chair in Biological Sciences and Chemistry at Goucher College, a small liberal arts college in Baltimore, Maryland. Dr. Segarra co-founded and co-directs the Alliance to Catalyze Change for Equity in STEM Success (ACCESS; NSF 1744098) organization, a group of scientific society diversity-focused committees that exchange promising practices and data. She is also co-primary investigator on the ADVANCE Partnership grant that sponsors the growth of ACCESS into a larger network of societies, known as ACCESS+ (NSF 2017953).

Simone B. Soso (PhD) is the Founder and Lead Consultant of STEM Futures Consulting, LLC and was previously Program Director of the Quality Education for Minorities Network. Dr. Soso guides the development, execution, and funding of professional development training events, technical assistance grant writing workshops, and communities of practice for faculty at MSIs. She is a trained Animal Ecologist with a BS in Animal Science from the University of Maryland Eastern Shore, and an MS in Animal Health Science from North Carolina Agricultural and Technical State University. She earned her PhD at Iowa State University in Environmental Science.

Contributors

Nancy Amador (MSCR) is a project manager at Kean University's School of Integrative Science and Technology. In her role, she works closely with undergraduate and graduate students across various STEM disciplines, providing resources and support to facilitate their transitions to advanced academic opportunities such as master's and PhD programs. Nancy has served as the primary contact for the North Region of the Computing Alliance of Hispanic-Serving Institutions, supporting institutions across the Northeast. She also collaborates with programs like the National Center for Women and Information Technology to coordinate events and initiatives that celebrate student achievements in computing. Nancy is an affiliate of the New Jersey STEM Pathways Network, an alliance dedicated to enhancing STEM education statewide from K–12 through higher education. She holds a BS in general biology from Kean University and an MS in Clinical Research (MSCR) from Drexel University. She has also earned a certification in teaching within inclusive STEM environments and is currently advancing her expertise in project management for university grants and research. In addition to her work at Kean University, Nancy volunteers as a program director at her local global church campus, ensuring seamless technology operations during Sunday services at a multi-site location.

Lisa Black (PhD) is the current Raising Equity Values with the Inclusive Professional Framework and ISO DEI Standards in Societies (REVIIS) co-principal investigator. After a career in social work, policy development, and leading strategic planning initiatives within the K–12 public education sector, in 2016 Black joined the staff of the American Society of Civil Engineers (ASCE), the nation's oldest engineering society, representing more than 160,000 members. In her role as ASCE's director of cultural belonging and social ethos, she is responsible for developing, monitoring, and evaluating policies, programs, activities, and products aimed at advancing a just, equitable, diverse, and inclusive civil engineering profession. Through her work with ASCE's board-level committee, Members of Society Advancing an Inclusive Culture (MOSAIC), she provides leadership in developing and implementing policies, programs, and products designed to advance awareness and recognition of the civil engineering profession, with a special emphasis on promoting civil engineering as an inclusive discipline. She holds a Bachelor of Arts in Sociology from North Carolina State University and both graduate degrees, a Master of Science in Social Work and PhD in educational leadership and policy analysis, from the University of Wisconsin, Madison. Dr. Black is also certified in the International Organization for Standardization's Diversity & Inclusion Standard (ISO 30415: 2021). She finds joy in reading scripture and spending time with her husband, their four adult children, and the many extended family members that life has blessed her with, including her chocolate lab, Odie.

Paula Brantner (JD) is an attorney experienced in harassment and discrimination law and the president and Principal of Accountability Ignited, a consulting firm in Silver Spring, Maryland. She works with scientific and professional societies to implement codes of conduct and improve meeting safety through reporting systems and allies programs.

Brian A. Burt is the current Raising Equity Values with the Inclusive Professional Framework and ISO DEI Standards in Societies (REVIIS) PI as well as Director and Chief Research Scientist at the Wisconsin Equity and Inclusion Laboratory.

Alison Cawood (PhD) led the conference planning teams for 2021, 2022, and 2023 Association for Advancing Participatory Sciences (AAPS) conferences, and she is currently a member of the AAPS Board of Directors.

Scott L. Collins (PhD) is Distinguished Professor of Biology at the University of New Mexico. His research focuses on the impacts of global environmental change on grassland ecosystems. He is the former editor-in-chief of *BioScience*.

Raena Cota is an experienced and dedicated professional in computer science education, committed to advancing diversity, equity, and inclusion in the tech sector. As the southwest connector for the Computing Alliance of Hispanic-Serving Institutions (CAHSI), Raena coordinates regional activities for CAHSI partners in Arizona, New Mexico, and Texas and organized bi-weekly meetings, workshops, seminars, and social events to foster collaboration and support within the HSI community. Beyond her work with CAHSI, Raena holds several leadership roles in the broader computer science-education community. She is the past secretary (2020–4) and current university liaison for the Computer Science Teachers' Association, New Mexico Chapter, contributing to statewide efforts to improve computer science education. Additionally, she is a founding member (2022–present) of the New Mexico Expanding Computing Education

Pathways Alliance, a statewide initiative focused on creating inclusive pathways for computing education in New Mexico. Previously, she served on the Academic Alliance Advisory Board for the National Center for Women and Information Technology (NCWiT) from 2020 to 2021 where she helped shape strategies for supporting women in computer science-education pathways and careers. Raena has also been the NCWiT Aspirations in Computing co-coordinator for the Southern New Mexico Region since 2014, playing a pivotal role in inspiring and supporting young women in their pursuit of computing careers. As a passionate advocate for diversity in tech, Raena regularly presents at national conferences, including the Research on Equity and Sustained Participation in Engineering, Computing, and Technology; Special Interest Group Computer Science Education; and Society for Information Technology and Teacher Education conferences, where she shares research, best practices, and strategies to sustain diverse participation in computing and technology fields.

DaJoie R. Croslan (PhD) is the chief operations and diversity officer at the American Institute of Biological Sciences (AIBS). She works closely with staff and stakeholder groups to identify and develop solutions to operational challenges, to promote AIBS programs, to cultivate relationships, and to ensure that AIBS is working internally and externally to foster a more diverse and equitable organization and society. She serves as the chair of the AIBS Diversity Committee and oversees all of AIBS's diversity initiatives.

Nohelia Diplan (MSW) is Managing Assistant Director at Kean University's department of computer science and technology, where she works closely with computer science and information technology students in their transition to new academic environments by providing support and education of the resources available for student success. She serves as the chair for the New Jersey Affiliate of the National Center for Women and Information Technology's Aspiration in Computing Program that focuses on increasing the representation of women, genderqueer, or nonbinary individuals in computing. Before joining Kean, she worked as a children and family therapist with children in the child-welfare system. In addition, Nohelia has experience working with a wide variety of mental health issues from youth to adulthood. Currently, she's pursuing a doctorate in education at Rutgers University, Graduate School of Education. Her area of research focuses on creating inclusive educational environments for students from historically excluded groups and understanding student experiences as they navigate through these transitions developing their professional identities. She holds a BA from Kean University in psychology and an MSW from New York University (NYU) in social work. While at NYU, she traveled and lived abroad in Buenos Aires, Argentina where she worked on community mental health outreach in a hospital setting and with the homeless population.

Kimberly Eck (MPH, PhD) is Senior Associate Vice President for Research at Emory University. In this role she established and oversees the Office of Research Development and the Training Grant Support Office, as well as the university's competitive intelligence function for the research ecosystem. She has been working in research development since 2009. Prior to joining Emory, Dr. Eck served as Assistant Vice Chancellor at the University of Tennessee, Knoxville; Associate Vice President at Geisinger Health System; the manager of collaborative proposal development at the Research Foundation for the State University of New York; and as a consultant at a small private consulting firm. She is the former president of the National Organization for Research Development Professionals (NORDP) and is the founding senior director of the NORDP Consultants Program, which seeks to expand the research ecosystem by providing research development services. She received her BA in sociology from the University of Notre Dame, MPH from Saint Louis University, and her PhD from the State University of New York, University at Albany, in epidemiology.

Sharday N. Ewell (PhD) is Assistant Professor of Inclusive STEM Pedagogy in the Department of Biology at the University of Mississippi in Oxford, Mississippi. Her research interests center on identifying strategies that improve learning for all and promote STEM persistence. Specifically, Dr. Ewell is especially interested in student approaches to learning and identifying strategies that improve their development for all students. Prior to starting at the University of Mississippi, she was a postdoctoral fellow in the Ballen lab at Auburn University. She was also a Teach for America corps member and middle school science teacher in North Carolina, Virginia, and Washington, DC. She is currently the co-chair of the PEER Committee for the Society for the Advancement of Biology Education Research.

Bethany Farmer (MSIS) is the community librarian at the Women in Engineering ProActive Network (WEPAN) and on the ACCESS+ initiative. She is dedicated to promoting diversity and inclusion in information representation and access.

As librarian for the ARC Network, Bethany has exponentially increased the size of the library, adding more than 7,000 resources related to intersectional diversity in the STEM fields. She has also developed a number of web guides to help those impacted by state and federal anti-DEI legislation. She serves as a team member on multiple ADVANCE Partnership grants.

Charles B. Fenster is the editor-in-chief of *BioScience* and Professor and Director of the Oak Lake Field Station at South Dakota State University. He is also past president of the American Institute of Biological Sciences, with a long history of service on the society's board of directors. Dr. Fenster's main research themes are the genetics of adaptation, the contribution of mutation to selection response, and quantifying pollinator-mediated selection on floral form and function.

Stephen A. Gallo (PhD) actively studies the biomedical grant review process in collaboration with the American Institute of Biological Sciences. He researches the intersection of reviewer and application characteristics, evaluation procedures, and their influence on peer review effectiveness.

Marcos E. García-Ojeda (PhD) is award-winning teaching Professor and a founding faculty member at the University of California (UC), Merced, where he has worked since 2006. In August 2024 he transitioned to the newly established Department of Medical Education. His research interests focus on how teaching spaces contribute to student learning and motivation as well as understanding the affective contributions of course-based undergraduate research experiences. Dr. García-Ojeda's contributions to teaching and faculty mentoring earned him the UC Merced Academic Senate Distinguished Undergraduate Teaching Award in 2012 and the Senate Award for Excellence in Faculty Mentorship in 2018. Previously, he served as Associate Dean of equity, justice, and inclusive excellence and is currently the chair for the Equity, Diversity, and Inclusion Academic Senate Committee. He is the past president for the Society for the Advancement of Biology Education Research, a national organization that promotes evidence-based teaching and learning in biology.

Anna Hiatt (PhD) is Assistant Professor of Biology at Hastings College in Hastings, Nebraska. Her PhD training is in undergraduate biology education–assessment development and she is highly experienced in teaching foundational biology courses. She was the primary investigator of the NSF-funded Advancing the Introductory Biology Experience, leads the National Association of Biology

Teachers (NABT) Introductory Biology Task Force, and has served in several leadership positions at NABT, including on the NABT Board of Directors. As a first-generation college student, she has a special interest in supporting fellow first-generation college students, faculty, and staff at her institution.

Lucas B. Hill (PhD) is an experienced researcher (III), evaluator, and organizational change consultant in the Wisconsin Center for Education Research (WCER) at the University of Wisconsin, Madison. He is Director of Research for the Center for the Integration of Research, Teaching, and Learning (CIRTL) and co-leads the SEER Institute. His work primarily focuses on understanding and assessing the effectiveness of complex, large-scale, and collaborative higher education reform, particularly related to the preparation of faculty, the adoption of evidence-based pedagogical practices, and the broadening participation of underrepresented students in STEM education. His portfolio includes several NSF-funded projects, including the NSF Eddie Bernice Johnson INCLUDES Aspire Alliance, the Faculty Inclusive Teaching Survey (FITS), the Inclusive STEM Teaching Project, Belonging in Field Education, and the CIRTL Change Leadership Development Program.

Abigail Hils is Senior Administration Officer for the United Nations Educational, Scientific and Cultural Organization at the Australian National Centre for the Public Awareness of Science. They served on the working groups for the 2021 and 2022 Association for Advancing Participatory Science conferences and led the conference planning team in 2022 before moving to the Australian Citizen Science Association.

Sarah Hug (PhD) is the director of Colorado Evaluation and Research Consulting (CERC). Her doctoral training in learning sciences and her background in K20 teaching informs her social science and evaluation efforts. CERC evaluates broadening participation efforts in engineering and computer science disciplines, and conducts social science research on equity, inclusive excellence, and student experiences in higher education.

Vedham Karpakakunjaram (PhD) is the Interim Dean of Chemical and Biological Sciences at Montgomery College, in Maryland. Dr. Karpakakunjaram has served as the Acting Department Chair at the Biology Department, Rockville Campus, Montgomery College during 2024–25. He is currently the President-Elect of the National Association of Biology Teachers (NABT). He is a former

Two-Year College Section chair (2021) of NABT and served on the President's Advisory Committee for Equity and Inclusion at Montgomery College. He was the co-principal investigator of Quantitative Biology at Community Colleges (NSF 1919613, 2019-25).

James Felton Keith is an engineer and economics researcher at Keith Institute. He is the founder of the Inclusion Score, author of the International Standards Organization ISO 30415: DISM (Diversity, Equity, Inclusion, and Belonging Service Management) methodology; and serves on the Raising Equity Values with the Inclusive Professional Framework and ISO DEI Standards in Societies (REVIIS) Project Team. He is also a co-principal investigator on the National Science Foundation's ADVANCE–REVIIS Partnership.

Jacklyn S. Kelty (PhD) is Chemical Hygiene Officer and Laser Safety Officer at San Jose State University. She is a co-principal investigator for the Knights Landing Environmental Health Project. She served as a co-chair of the Association for Advancing Participatory Science's Justice, Equity, Diversity, and Inclusion Working Group from 2020 to 2023.

Stacey Kiser is Biology Instructor and Undergraduate Research Faculty Coordinator at Lane Community College in Eugene, Oregon. Stacey is a past president of the National Association of Biology Teachers (2014).

Gretalyn (Gretal) **Leibnitz** (PhD) is an experimental psychologist; executive director of ProActualize Consulting, LLC; and an associate faculty member of the College of Engineering and Computer Science at the University of Texas, Rio Grande Valley. Dr. Leibnitz works to identify, develop, and apply evidence-based strategies to promote systemic change to advance organizational and disciplinary excellence. Dr. Leibnitz is the current co-principal investigator (PI) and project director of Raising Equity Values with the Inclusive Professional Framework and ISO DEI Standards in Societies (REVIIS); current co-PI and project lead on the Latina THRIVE Partnership; the past ACCESS+ project PI; and the founding and current director of the ADVANCE Implementation Mentors (AIM) Network.

Melanie Martin (PhD) is Professor of Computer Science at California State University (CSU), Stanislaus. Her research interests include natural language processing, team communication, data science, health informatics, and computer science and mathematics education. Dr. Martin works to increase diversity in STEM fields and to provide opportunities and paths to success for students by using and developing evidence-based methodologies. Dr. Martin has served as co-PI for two NSF Scholarships in STEM grants (NSF 1643944, 2221680). In addition, Dr. Martin served as the campus program director and coordinator for the CSU Louis Stokes Alliance for Minority Participation for twelve years. At CSU Stanislaus, she is currently the Computing Alliance of Hispanic-Serving Institutions campus lead, the Cal-Bridge campus liaison, and a research mentor for undergraduate students.

Sherry Marts (PhD) is a former scientific society chief executive officer with a wide-ranging background in biomedical research, nonprofit management, public education, and research advocacy. Her interest in the issues of harassment and bullying lies at the intersection of her professional life as a woman in science and her experience as a women's empowerment self-defense instructor.

Heather E. Metcalf (PhD) is Managing Director of Research, Innovation, and Entrepreneurship at the Women in Engineering ProActive Network (WEPAN). Dr. Metcalf holds a BS and MS in computer science, a BS in applied mathematics, an MA in gender studies, and PhD in higher education, science, and technology policy. Their applied work bridges research, policy, and practice to advance intersectional gender equity and systemic change in STEM higher education and professional settings. Dr. Metcalf serves as an investigator on several private and federally funded projects, including their role as co-PI for ACCESS+, an NSF-funded initiative (NSF 2017953) that engages STEM professional societies as sites of transformation.

Kirstin J. Milks (PhD) learns with and from AP® Biology and introductory science students at Bloomington High School South in Bloomington, Indiana. She is a National Board–certified teacher, a Presidential Awardee for Excellence in Math and Science Teaching, and a senior fellow at the Knowles Teacher Initiative. Dr. Milks currently serves as book review coeditor for the *American Biology Teacher* and is the 2025 president of the National Association of Biology Teachers.

Nicola Mohan (MSW) is the diversity, equity, inclusion, and belonging (DEI&B) senior manager for the Society of Women Engineers (SWE). Mohan's work spans various internal and external DEI&B initiatives at SWE. Originally from

Trinidad and Tobago, Mohan migrated to the United States as an undocumented immigrant at the age of three, a personal journey that fuels her ongoing advocacy for DEI&B initiatives. Possessing a bachelor's degree in political science from the State University of New York at Geneseo, a master's degree in social work from Boston University, and a Diversity and Inclusion for HR certificate from Cornell University, Mohan brings a comprehensive educational background to her impactful role at SWE.

Stephani Page (PhD) serves as Managing Director of programs, strategic initiatives, and community engagement for the Women in Engineering ProActive Network (WEPAN), a nonprofit centered on advancing cultures of inclusion and equity in engineering education and the engineering workforce since 1990. She also serves as co-PI on WEPAN's NSF ADVANCE–funded ARC Network and ACCESS+ initiatives as well as the NSF INCLUDES-funded Collaborative Change Consortia the Women of Color in Engineering Collaborative. She is a STEM equity professional who has nearly twenty years of interdisciplinary academic research experience. She earned her doctorate in Biochemistry and Biophysics from the School of Medicine at the University of North Carolina at Chapel Hill. Dr. Page is a proud graduate of North Carolina A&T State University with a bachelor's in chemical engineering and master's in biology. Her research has been recognized by the Biophysical Society, the American Society for Biochemistry and Molecular Biology, and the American Heart Association. Throughout her career, Dr. Page has been committed to diversity, equity, and inclusion in the academy and STEM. She is the creator of the #BLACKandSTEM community which has a social media following of nearly 20,000. For nearly twenty years, she has contributed to several major efforts around equity in STEM working with entities such as the National Institutes of Health, the National Science Board, and the National Academies of Sciences, Engineering, and Medicine. Dr. Page has been recognized in mainstream outlets including *Fast Company*, *NPR*, *Nature*, and *Forbes*.

Samiksha Raut (PhD) is Associate Professor in the Department of Biology at the University of Alabama at Birmingham (UAB). Her research interests in biology education are focused on analyzing the factors that promote teaching and learning in the context of peer-to-peer mentoring groups for gateway courses. She has an active area of interest in using service-learning as a community engagement tool to educate non-STEM majors. Dr. Raut's teaching and mentoring efforts have been widely recognized by various state- and national-level awards. She is the co-founder of the ROSE (Research on STEM Education) network at UAB and the Society for the Advancement of Biology Education Research PEER Network. She currently serves as an editorial board member to *CourseSource, Frontiers in STEM Education,* and the *Journal of Microbiology and Biology Education.*

Jaclyn Reeves-Pepin has served as Executive Director of the National Association of Biology Teachers (NABT) since 2011. In addition to overseeing NABT operations and strategic initiatives, she is also the lead coordinator of NABT programs where she creates new professional learning resources and opportunities for K–16 biology and life science educators.

Roberta Rincon (PhD) is Director of Research and Impact for the Society of Women Engineers. Dr. Rincon oversees the organization's research activities on equity issues that affect girls and women in engineering and serves as the PI for the Women of Color in Engineering Collaborative.

Miriam Segura (PhD) is Biology Professor and the Harry B. Forester Eminent Scholars Chair in the Department of Biology, University of North Georgia (UNG) in Dahlonega, Georgia. Her research interests focus on the analysis of primary scientific literature in the classroom and on initiatives that help diversify the membership and leadership of professional societies. In 2017 she was named one of the Inspiring Leaders in STEM by *INSIGHT Into Diversity Magazine.* Dr. Segura's teaching was recognized in 2016 through UNG's most prestigious honor, the Distinguished Teaching Award, and in 2017 through the University System of Georgia Regents' Teaching Excellence Award (Felton Jenkins, Jr. Hall of Fame Faculty Award). She is the past-president for the Society for the Advancement of Biology Education Research (SABER), and she is the co-founder of the SABER PEER Network. Dr. Segura has served as an editor for the *Journal for Microbiology and Biology Education* and the STEM Education Section of *Frontiers Journals.*

Learty L. Shaw (PhD) is principal consultant and founder of Right Research and is a consultant for the ACCESS+ initiative at the Women in Engineering ProActive Network. He also serves as the director of institutional research and decision support at Mercer University and holds a research faculty appointment in the Tift College of Education.

Linda Silka (PhD) is Psychology Professor at the University of Maine and a senior fellow at the Mitchell Center for Sustainability Solutions. She served on the working groups for the 2012, 2015, 2017, 2019, 2021, and 2022 Association for Advancing Participatory Science conferences.

Ershela L. Sims (PhD) is Executive Director and CEO of the Women in Engineering ProActive Network (WEPAN). She is the PI for WEPAN's ACCESS+ initiative and co-PI on a number of NSF-funded projects including the ARC Network, KnowlEDGE, and the National Center for Women and Information Technology (NCWiT) ADVANCE Partnership grant. She is also a member of the board of directors for the NSF INCLUDES Women of Color in Engineering Collaborative (WCEC) as well as the Executive Steering Board for the NSF ADVANCE-funded THRIVE Partnership. Dr. Sims has over twenty-seven years of experience broadening participation in STEM ranging from mentoring, training teaching fellows on inclusive practices through interactive theater, and leading employee resource groups, to serving as a faculty adviser, regional leader, and boundary spanner for a number of STEM professional societies. Her career in STEM has included work in industry, higher education academia and research, and K–12 engineering education and administration. Dr. Sims holds a BSE in biomedical engineering from Duke University, a PhD in biomedical engineering from the University of North Carolina, and a certificate in Leading with Diversity Equity, and Inclusion from the University of Utah. Her work on broadening participation in STEM has been featured on JJ DiGeronimo's *Tech Savvy Woman* "5 Minutes with NC STEM State Leader & STEM Award Winner" as well as *PBS Black Issues Forum* "Step Up STEM."

James M. Verdier is the senior editor of *BioScience*, the journal of the American Institute of Biological Sciences (AIBS). In addition, he works with AIBS staff on programs aimed at increasing diversity and inclusion in the life sciences, including those focused on professional education and science communication.

Lydia Villa-Komaroff (PhD) is a molecular biologist, a businesswoman, and a diversity advocate. After failing retirement twice, she founded Intersections Consulting as an umbrella for activities that include business boards, not-for-profit boards, and consulting with academic institutions, professional societies, government agencies, companies, and individuals. During her research career, she was among the early cadre of scientists to join DNA from animals or people to DNA from bacteria to make useful proteins and to study biological processes and human diseases. Dr. Villa-Komaroff has been a faculty member at the University of Massachusetts Medical Center, Children's Hospital (Boston), and Harvard Medical School. She was Vice President for Research at Northwestern University in Illinois and the Whitehead Institute in Cambridge and chief scientific officer and chief executive officer of Cytonome, a company that builds purpose-built cell sorters where she remains on the board of directors. She served on the boards of Transkaryotic Therapies and the American Type Culture Collection (ATCC). She is among the founding members of the Society for the Advancement of Chicanos/Hispanics and Native Americans in Science (SACNAS). She is a third-generation Mexican American and a descendant of the Spanish conquistadores. She earned her BA in biology from Goucher College in Baltimore and her doctorate from the Massachusetts Institute of Technology. Dr. Villa-Komaroff served as chair of the American Society for Cell Biology Minority Affairs Committee from 2005 to 2007, continuing the legacy Dr. George Langford started in the 1980s. She is an advocate for a diverse scientific enterprise that embodies a broad and humane perspective of science. She is a role model; her career demonstrates not only that race and gender barriers can be surpassed in STEM but also that scientists can be advocates for social justice without compromising scientific excellence.

Preface

Verónica A. Segarra, Mercy Mugo, and Simone B. Soso

In envisioning this book before 2025, our collective goal was to explore and highlight the roles scientific professional societies play in fostering diversity, equity, and inclusion within their disciplinary domains. This project germinated from a series of think tank sessions dedicated to better understand society inclusivity efforts. These discussions, rich in ideas and collaborative spirit, set the stage for a more profound investigation into the programmatic and policy strides being made to not only invite but also integrate a wider range of voices into the scientific community.

As this book of society case studies comes to fruition, we find ourselves in a markedly different political climate than when we began. The changing landscapes of policy and public opinion have put organizations to the test, compelling them to reassess their commitment to the values of diversity, equity, and inclusion. It is within this context that the stories and strategies outlined in this book gain even greater relevance. The chapters that follow are more than just case studies of intention; they are testimonies to the steadfastness with which these societies are navigating the complexities of modern governance and public accountability in their pursuit to uphold these essential values.

The societies whose efforts grace these pages exemplify what it means to espouse values and act on them decisively. Each chapter details the concrete steps these organizations have taken to ensure that all scientists, especially those who have historically been marginalized, are represented and heard at the decision-making table. As readers engage with this collective narrative, we hope that the work documented here inspires renewed dedication, reflection, and deeper dialogues, nurturing pathways toward a more inclusive and equitable scientific community.

Acknowledgments

Building Inclusive Scientific Communities and Leadership: Case Studies from Professional Societies shares case studies of science-focused professional organizations and societies as they strive to make their disciplines inclusive through leadership and organizational change. We are indebted to the many author contributors and organizations who eagerly responded to our national call for chapters. We thank them for their willingness to share their stories—including their successes and challenges—as they have created, implemented, or changed programs, policies, and practices. For us this book has been a coming together of the minds, allowing many of us to learn from each other, deepen our relationships, and make new connections.

We thank our copyeditor, Ada Hagan (Alliance SciComm and Consulting, LLC), for her help in realizing our vision for a book that honored the efforts and voices of the author contributors and organizations represented. With her help, we ensured that our collective product had the necessary structure and storyline to best convey our thoughts.

This book has been organized by the Leveraging, Enhancing, and Developing Biology (LED-BIO) Research Coordination Network (RCN). The LED-BIO RCN was established with funding from the National Science Foundation (NSF) and its BIO-LEAPS (Leading Culture Change Through Professional Societies of Biology) mechanism (NSF 2134725). LED-BIO is a partnership that includes the Alliance to Catalyze Change in STEM Success (NSF 1744098), the Quality Education for Minorities Network, MSI STEM Research and Development Consortium, and the NSF INCLUDES ASPIRE Alliance (NSF 1834518, 1834522, 1834510, 1834513, 1834526, and 1834521).

Part One

Organizations Nurturing a Diverse and Inclusive Group of Visible Thought Leaders

Introduction to Part I

Organizations can champion equity and drive innovations by creating opportunities where professionals are nurtured and diversity is prioritized in their leadership ecosystems. The case studies in this part collectively emphasize the diverse approaches and initiatives aimed at advancing diversity, equity, and inclusion (DEI) within professional communities. They serve as practical examples of how intentional programming, leadership, allyship, and standardized inclusive practices can create environments where individuals feel valued and supported.

There are five case studies in this part. The first is about a network that seeks to foster a sense of belonging and create a supportive environment for individuals from historically excluded groups through varied programming. The second study highlights the importance of structured and intentional efforts to promote student engagement and professional development. The third case study examines the role of allies programs in fostering safety and inclusiveness at scientific meetings and conferences. The fourth study provides insights into the strategies and challenges experienced by a STEM (science, technology, engineering, and mathematics) society leader in promoting DEI. The last study emphasizes the importance of a framework designed to ensure equitable access to opportunities and resources. By examining these real-world instances, readers can gain actionable insights into effective practices and strategies for fostering inclusivity and addressing systemic barriers in various contexts.

1

The PEER Network: A Catalyst for Inclusive Leadership in the Society for the Advancement of Biology Education Research

Miriam Segura, Marcos E. García-Ojeda,
Samiksha Raut, and Sharday N. Ewell

The lack of diversity and equity in science, technology, engineering, and mathematics (STEM) fields is pervasive (NCSES 2021; Pearson et al. 2022; Popo-Olaniyan et al. 2022; Trapani and Hale 2019), with only about 10 percent of the STEM workforce being from historically excluded groups (NSB 2015; NSF 2019; PCAST 2012).[1] This lack of diversity extends to the field of biology education research (BER) (Segura-Totten et al. 2021). Research shows that diverse teams can solve problems and advance disciplinary research better when compared to non-diverse groups (Borman et al. 2010; McGee 2020a; Page 2007, 2017). In BER, this lack of diversity may lead to the framing of educational problems and solutions from the skewed perspective of the dominant white culture, which misses important points of view that enrich the development of evidence-based educational practices and how they apply to PEER (persons excluded due to their ethnicity or race) (Asai 2020) students (Thompson et al. 2020).

Professional societies serve as hubs for researchers to network, share, and advance their work (Hulede 2018; Primus et al. 2022; Womack et al. 2020). Exclusionary environments may alienate individuals from historically excluded groups, blocking the social capital built during these meetings and negatively impacting their careers (Skvoretz et al. 2020). Several initiatives—such as financial support in the form of travel awards, conference organizers' considerations of meeting affordability, location, and accessibility as well as speaker and planning committee diversity—have helped address the lack of professional society diversity (Leibnitz et al. 2022a; NASEM 2023; Segarra et al. 2020a; Skvoretz et al. 2020). However, there has been little emphasis on processes to diversify the leadership and governance structures of professional societies in STEM-related fields (Segarra and Etson 2023; Yen 2019). Recently, the Society for the Advancement of Biology Education Research (SABER) took steps that helped to change the culture of the society. This case study aims to describe initiatives that led to a more inclusive environment, the pathways to diverse leadership that these initiatives created, and the challenges, obstacles, and opportunities encountered during the process of diversifying SABER so that other professional societies may benefit from our experiences.

Author Positionality

The authors identify as Black, Latinx, Southeast Asian, cisgender men and women, and LGBTQ+ (lesbian, gay, bisexual, transgender, queer/questioning) persons. Some authors have been involved with the governance of SABER at the highest levels and others have been involved in SABER committees and special interest groups and as conference attendees. Our positionality as members of SABER and individuals from historically excluded communities informs the lens through which we view the process of diversifying the society. Our positionality also continues to drive our involvement in the diversification of SABER.

Research Questions

This case study aims to explore two questions:

1. To what extent does a member-initiated group effect institutional change in a professional society?
2. What are the challenges faced, and lessons learned, by the PEER Network leadership in its journey to change the culture of SABER?

Background for the Study

SABER was founded in 2010 and held its first annual meeting in 2011 (Offerdahl et al. 2011). Since then, it has become the preeminent organization for research on undergraduate biology education. Although SABER was founded on principles of fiscal inclusion, most of those who created the society and held leadership positions for the first decade were white presenting, which led to a predominantly white organizational culture. Data from 2020, the first year that the society collected demographic indicators, suggest that PEERs are highly underrepresented (Segura-Totten et al. 2021). In this article, we use the term "PEERs" to include individuals who identify as members of racial and ethnic groups that have been traditionally excluded, even if not traditionally underrepresented, in STEM. This is because certain racial and ethnic groups still experience exclusion in science even if they are numerically well represented (Siy and Cheryan 2013; Yip et al. 2021). A post-conference survey in 2019 revealed that many SABER members felt there was a lack of racial diversity at the annual meeting and in the organization's leadership, and that first-time meeting attendees did not feel included (Segura-Totten et al. 2021). Because building a critical mass of historically excluded individuals reduces negative feelings and stereotypes, increases their full participation, and leads to improved self-confidence and ambitious goal setting (Asgari et al. 2010; Barabino et al. 2023; Inzlicht and Ben-Zeev 2003; Sekaquaptewa and Thompson 2002; Stout et al. 2011), growing PEER membership in SABER became an important priority.

With the murder of Mr. George Floyd in 2020, SABER began a process of self-reflection on its practices of inclusion and diversity. As a result, the Diversity and Inclusion Action Group on Place and Racial Justice was created. This group was open to all SABER members. The self-reflection process included a self-study of the organization during the 2020 annual meeting, which led to suggestions of increasing representation of and building community for PEERs in SABER, and awareness of diversity, equity, and inclusion (DEI) issues within the society (Segura-Totten et al. 2021). In response to these recommendations and as a result of conversations within the Diversity and Inclusion Action Group on Place and Racial Justice, three members of the racial justice team volunteered to initiate a special interest group (SIG), named the PEER Network, to increase the belonging of PEERs in SABER. This SIG was created as a year-round community that meets in person during the SABER annual meeting and virtually the rest of the year, which provides a consistent space for individuals to grow professionally and build community. The first organizational meetings of the SIG occurred during the 2020 SABER annual meeting.

Theoretical Frameworks

Although the effect of communities like the PEER Network in professional societies is understudied (Barabino et al. 2023; Leibnitz et al. 2022a), frameworks suggest that networking with peers and building relationships with role models from historically excluded backgrounds in high-status positions can act as a social vaccine that enhances individuals' fit in their local environment and protects them against negative stereotypes (Dasgupta 2011; Stout et al. 2011; Thiem and Dasgupta 2022).

This case study draws from the social inoculation model (Dasgupta 2011) and the inclusive professional framework for societies (IPF: Societies) (Leibnitz et al. 2022a). The social inoculation model posits that having a critical mass of individuals from a certain background in the ingroup of an organization inoculates others of the same backgrounds against issues of social belonging and negative stereotypes (Dasgupta 2011; Stout et al. 2011). The social inoculation model does not suggest adapting historically excluded individuals to the dominant culture's environment. Instead, it proposes ways to highlight role models and communities that align with the self-concept of historically excluded individuals to change the environment (Dasgupta 2011). The IPF: Societies framework focuses on four key factors for organizational culture change: identity awareness, intercultural mindfulness, inclusive relationships, and influential DEI actions (Leibnitz et al. 2022a). It suggests that when individuals with an equity mindset—encompassing identity awareness and intercultural mindfulness—collaborate on equity-focused initiatives, they can foster inclusive relationships and drive influential actions that catalyze cultural change within an organization (Leibnitz et al. 2022a). The IPF: Societies framework also identifies DEI change leaders who connect ideas and stakeholders to promote cultural change (Leibnitz et al. 2022a). In this case study, we will describe the initiatives to diversify SABER and how they helped to drive culture change as explained by the above theoretical frameworks.

Case Study Narrative

The creation of the PEER Network was an important first step for increasing inclusion and amplifying the voices of PEERs in SABER. The following section details how we built the PEER Network, including how to design a space focused on member priorities, organizational approaches that allowed us to divide up the workload equitably, communication tools that allowed the network to thrive, lessons learned, challenges encountered, and opportunities discovered.

Initial Organization of the PEER Network

Following initial PEER Network meetings, the founding members met to plan the next steps. To help shape the direction of the group and to collect contact information for individuals interested in joining, we designed a survey and distributed it using the SABER email listserv. The results of this survey showed that individuals were most interested in participating in professional development opportunities (like journal clubs on articles related to DEI), mentoring groups, and social opportunities (like check-ins and regular meetings to discuss topics of importance to PEERs). Other priorities were increasing the visibility of PEERs within SABER, building research collaborations, mental health resources, a space to discuss issues of importance to PEERs, fundraising, and the creation of a group listserv. We addressed the priorities identified through the survey by creating five subgroups, which were led by member volunteers.

1. Happy Hour, which focused on providing social activities and community building.
2. Journal Club, which focused on the discussion of primary research articles relevant to PEERs.
3. Safe Space, which provided a forum to discuss issues and concerns and was the only subgroup restricted to PEERs.
4. Mentoring, which followed the existing SABER model where one senior mentor is paired with several mentees based on the mentee's needs and geography.
5. Advocacy, which focused on raising the visibility of PEERs in SABER and advocating for their needs to the SABER leadership.

These subgroups met several times in the fall, spring, and summer. Importantly, a community's needs are shaped by the environmental and social conditions of the time. For instance, we established the PEER Network during the Covid-19 pandemic, shortly after George Floyd's murder, which likely heightened the demand for mental health and social support resources. To address the evolving needs, we assessed the group's priorities two to three times a year for the first two years, and then annually, to ensure that our initiatives remained member centered. We also received emails from non-PEER individuals wanting to join the group to support the diversification of SABER, which prompted us to consider the role of allies in the SIG. We decided to welcome allies and publicized this through emails to the SABER listserv and during group meetings. The inclusion of allies in the group is an important component, since the burden of changing the organizational culture should not exclusively fall on members from historically excluded groups (Li et al. 2023).

PEER Network members met as a group in the fall and spring from 2020 to 2023. Meetings started with introductions, followed by discussions on the future of the group and the initiatives that mattered most to members. We then provided a description of the different subgroups to encourage individuals to either join or help lead those that matched their interests. The meetings ended with a community-building activity and a request for attendees to spread the word about the SIG and invite others who might be interested. Because we also wanted the group to serve as a recruitment tool to increase the diversity of the SABER membership, we asked participants to reach out to individuals regardless of whether they were SABER members. We also publicized the group on listservs outside the organization, such as those for the Council for Undergraduate Research and the American Society for Microbiology Biology Scholars Program.

Gaining Momentum and Establishing an Online Presence

As the initial subgroups began meeting, the Writing Buddies subgroup—which paired individuals interested in meeting once or twice a week to work on research-related projects—was formed in response to member needs. Another member-driven priority was to establish an online presence within SABER to increase the visibility of PEERs, attract new network members, and facilitate communication. To accomplish this, we solicited volunteers to act as web-page liaisons between PEER Network and the SABER website committee to build a web page for the group. The web page (https://saberbio.wildapricot.org/PEER-Network) contains a description of the goals and mission of the group, an announcement section, information on group initiatives, a community discussion forum, and contact information for joining the PEER Network. The design and content for the web page were refined using member feedback.

Information Flow, Communication Tools, and the Role of Technology in Creating an Inclusive Community

We felt that an important component in establishing a supportive community is effective communication to help align participants' values and foster a sense of community. Despite the challenges SABER members faced during the establishment of the PEER Network, our reliance on technology to connect during the pandemic likely strengthened our communication efforts. Besides meeting through Zoom, members used tools such as Slack, Google Drive (which served as a file repository), Google Spreadsheets, an email listserv, and a discussion board linked to the PEER Network page within the SABER website. Of these, Slack and the SABER website discussion board were the least successful in mediating communication. At the time, Slack was a fairly new technology, and unfamiliarity with the platform as well as the institutional use of competing products (like Microsoft Teams) likely contributed to its lack of use. The SABER discussion board is only available to active society members, which restricted access and potentially limited its effectiveness. On the other hand, the accessibility of the SABER email listserv (which is not restricted to members) and the widespread adoption of the Google platform probably explain why these tools were used more extensively.

Information in the group flowed from the SIG chairs to the subgroup leads and then to subgroup members. The chairs communicated directly with the subgroup leads to track the progress and success of initiatives, and they also sent announcements through group-wide emails. The subgroup leads solicited participants, made announcements through the listserv, and communicated with subgroup members via email. Group members met two to three times in person during the SABER annual meeting to discuss progress made, set priorities for the following year, and network and socialize.

Currently, the PEER Network has almost 200 members. The creation of this group provided fertile ground for important discussions related to DEI within the society. Group meetings offered a way for PEERs to connect and share their experiences. Most importantly, this group gave PEERs a collective voice within SABER that had previously been marginalized or overlooked, and it created a community for like-minded professionals from historically excluded racial and ethnic backgrounds (Harte et al. 2009; Zambrana et al. 2015).

Data Collection and Analysis

To determine the demographics of the PEER Network and the impact of the group on members, we designed and deployed a survey with Likert scale, multiple select, and open-ended questions. Descriptive statistics were obtained using Microsoft Excel. Survey participants gave their informed consent to participate in the study (IRB #2021-123, University of North Georgia). We used ChatGPT-4o to suggest peer-reviewed references and assist with revising grammar in this manuscript.

Findings

PEER Network: Promoting Inclusion and Raising Marginalized Voices

We surveyed the members of the PEER Network in 2022 to determine the initial impact of the community and to investigate the demographics of the group. Results ($n = 22$) indicated that 86 percent of respondents self-identified as PEERs and 27 percent are employed at historically Black colleges and universities, Hispanic-serving institutions, and minority-serving institutions (MSIs). Notably, 64 percent of respondents had been part of the PEER Network for a year or longer, 32 percent had been members for 6–11 months, and 4.5 percent had belonged to the group for less than 6 months. This suggests that while the group has provided community and networking opportunities to participants, we have not attracted many new members since the early days of the network. Additionally, 68 percent of survey respondents were relatively new to SABER, having attended the annual meeting three times or less. This aligns with the 2020 finding that most PEERs in SABER at the time had attended only one meeting (Segura-Totten et al. 2021). To better understand the usefulness of PEER Network components, we asked respondents about their participation in group activities. The Writing Buddies and Journal Club subgroups were the most popular activities among respondents. Additionally, participants expressed interest in engaging in advocacy and mentoring initiatives in the future. These responses indicate that group members are particularly drawn to initiatives that support career advancement and amplify their voices within the organization.

Table 1.1 Results of a Survey of the PEER Network Membership

Prompt	Mean ± SEM
The PEER Network has affected my overall perception of SABER.	3.59 ± 0.14
The PEER Network has affected my perception of inclusion and diversity efforts in SABER.	3.41 ± 0.14
As a result of my participation in the PEER Network, my relationships with other members of SABER have been affected.	3.55 ± 0.14
As a result of my participation in the PEER Network, my sense of belonging in SABER has been affected.	3.45 ± 0.14
As a result of my participation in the PEER Network, the likelihood that I will remain a member of SABER (or that I will join SABER) has been affected.	3.55 ± 0.13

Note: Participants responded to prompts using a Likert-scale of 1, strongly negatively, to 4, strongly positively ($n = 22$). Despite this being a small sample size, the number of respondents represents 50–60 percent of the active PEER Network population at the time (around 100 members).

We also wanted to determine whether the SIG was meeting its goals of increasing the sense of belonging of PEERs in SABER and increasing the focus of the society on DEI. The mean responses for all prompts were in the "positively" to "strongly positively" range (i.e., numerical scores between 3 and 4; Table 1.1). The survey results suggest that the SIG has been successful in improving participants' perception of SABER and its DEI efforts, positively influencing SIG members' relationships with others in the society and increasing SIG members' sense of belonging in SABER (Table 1.1). The results also suggest that participation in the PEER Network may increase the retention of its members within the organization (Table 1.1).

PEER Network: A Pathway to Leadership

The formation of the PEER Network led to the diversification of the SABER leadership: from 2020 to 2023, the demographics of the elected offices in the executive committee (EC) changed from 20 percent to 80 percent PEERs. The SABER EC is composed of the president elect, president, past president, treasurer, secretary, and the managing director (ex officio). The president serves a three-year term, with one year as president-elect, one as acting president, and the third as past president. The treasurer and secretary are elected to three-year terms, and the managing director is a permanent, half-time, staff position. All the PEERs elected to the EC from 2020 to 2023 were members of the PEER Network and four out of five obtained leadership experience through the SIG that enabled them to successfully run for office. We propose that this increased diversity within the organization's top leadership will encourage other PEERs to consider running for elected office. The social inoculation model further suggests that the exposure of individuals with limited experience in a field to high-profile peers from similar backgrounds—like those serving on the SABER EC—will lessen their feelings of self-doubt and challenge negative stereotypes, potentially inspiring them to pursue leadership roles.

Challenges

Because the PEER Network is run entirely by volunteers, an ongoing challenge is turnover within the group and how to facilitate continuity to maintain the community's momentum. After three years, some group leads experienced burnout due to the large workload. Several of the subgroup leads, as well as the chairs, stepped down and were replaced. To sustain the momentum of the group, one of the chairs made herself available to the new chairs in a mentoring capacity by establishing regular meeting times to address issues and clarify questions about running the group. This support has now been institutionalized through a change in the leadership structure for the network: chairs now serve a two-year term, with the first as incoming chair and the second year as chair. This allows the current chair to mentor the incoming chair for a year and facilitates continuity in the group's leadership. Around the same time, PEER Network transitioned from a SIG to a SABER committee (described further in the upcoming section "A seat at the table: The transition from SIG to committee"), which involved a structural rearrangement of the group and a shift of responsibilities for the chairs and subgroup leads. For these reasons, progress on group initiatives slowed compared to the initial years. While maintaining momentum remains an ongoing challenge, we found that transitioning PEER Network from a SIG to a committee has helped. The committee's direct reporting to the SABER EC and its monthly meetings with other committee chairs as well as the EC during steering committee meetings contribute to sustained progress and accountability.

Another challenge we encountered was catalyzing research collaborations between PEER Network members. In the spring of 2021, we attempted to jumpstart collaborations by asking those interested in establishing new research connections to present a short elevator pitch on their research project. Five individuals gave presentations via Zoom, followed by breakout rooms that were set up for people interested in collaborating with the presenters. Although the event was successful in building community between PEER Network members, to our knowledge, no research collaborations have yet resulted from it, which indicates that bringing members together alone is insufficient to promote collaborative work. Therefore, it is worth pursuing other avenues to incentivize research collaborations, like in-person networking events during SABER annual meetings or providing small seed grants to fund collaborative projects. In-person networking events during the SABER annual meeting would allow individuals more focused time to learn about each other's research projects and to plan collaborations, as the annual meeting offers a four-day window during which to meet, plan, and reflect.

A significant challenge that must be addressed is the persistent resistance to organizational change that often manifests as pushback from members who are entrenched in the status quo. This opposition not only hampers progress but also undermines efforts to create a more inclusive and dynamic environment within the society. One recommendation from the 2020 SABER self-reflection was to strengthen connections with the community surrounding the annual meeting's venue, the city of Minneapolis, Minnesota. In response to this recommendation, the 2024 meeting featured a location-based keynote speaker focused on issues of social justice with a special emphasis on education in Minneapolis. This decision sparked pushback from some SABER members, who perceived the initiative as compromising the academic rigor of the conference. However, both informal and formal feedback from attendees (through conversations and the post-conference attendee survey, respectively) was positive after the place-based keynote event. The feedback suggested that explicitly communicating the rationale for the place-based keynote speaker and clarifying that this activity was in addition to research-based keynote seminars at the annual meeting are important to securing buy-in from the SABER membership. Gaining buy-in from all stakeholders is crucial to reducing resistance and fostering a shared vision for the organization's future. To address this challenge, open and honest communication between stakeholders is essential, with the goal of finding a solution that respects and addresses the concerns of all members while still upholding SABER's commitment to inclusion and equity.

Opportunities

The creation of the PEER Network provided a space where individuals with similar goals and priorities could interact and exchange ideas, which led to new projects that continue to change the culture of SABER. For example, a group of PEER Network members submitted a grant proposal to the National Science Foundation BIO-LEAPS program. This proposal aims to enact sustainable culture change in SABER by leveraging the success of the PEER Network to continue to shift SABER's culture through (a) expanding the reach of SABER to individuals currently unaware of its existence; (b) providing funding, mentorship, and networking to PEERs and faculty at MSIs to sustain the involvement of diverse members within the organization; and (c) delivering inclusion-related mentorship to individuals in the SABER leadership. Additionally, the funding of this proposal in the spring of 2023 infused the PEER Network with much-needed funds, enthusiasm, and buy-in from the SABER leadership and membership. We hope that this renewed enthusiasm maintains the momentum of initiatives begun in 2020.

The PEER Network Advocacy subgroup created PEER Spotlights, interviews of PEER Network members who are actively involved in the organization. These increased the visibility of PEERs in SABER, since they are featured on the SABER website and are shared with the membership at large. Finally, more diverse leadership in the EC and SABER committees led to discussions of organizational priorities and initiatives that included the points of view of members from different ethnic and racial groups. This enriched conversations and resulted in innovative ways of moving the organization forward.

A Seat at the Table: The Transition from SIG to Committee

In 2023 the SABER EC decided to transition PEER Network from a SIG to a formal SABER committee involved in the governance of the organization. Committees do the work of SABER as an organization, while SIGs simply connect people with similar interests. For example, committee chairs attend monthly steering committee meetings, where the EC discusses and plans for new initiatives with the help of this broader group involved in SABER governance. Because the PEER Network initiatives directly benefited SABER members, the EC felt that the SIG should become a committee. Now a committee lead, one of the PEER co-chairs attends the monthly SABER Steering Committee meetings where they have a voice in shaping the future of SABER.

Becoming a committee led to the reorganization of the PEER Network in a couple of ways. First, committees have a limited membership, while SIGs have open membership. This meant that the PEER committee would be comprised of the co-chairs and the SIG subgroup leads. Second, initiatives that once ran internally to the SIG would now be advertised broadly through the SABER listserv and members email list, which leads to a higher visibility of initiatives to increase inclusion. These changes meant that while the core work of the PEER committee would be limited to a smaller number of individuals, the committee would offer the same suite of activities as the SIG but with a broader reach.

Discussion and Reflection

The PEER Network Case Study through the Lens of the Social Inoculation and IPF: Societies Frameworks

PEER Network has been successful at elevating the voices of PEERs, providing a pathway to leadership for members of the group, and increasing the sense of belonging of marginalized individuals. The success of the group can be explained by viewing the related initiatives and actions through the lens of the social inoculation and IPF: Societies frameworks (Table 1.2). As the IPF: Societies model predicts (Leibnitz et al. 2022a), an equity mindset was created in SABER's PEER Network through inclusive relationships nurtured by activities that build community. As suggested by the social inoculation model (Dasgupta 2011), activities were aligned with the self-concept of historically excluded individuals instead of adapting them to fit the current environment. The equity mindset created led to influential actions, such as travel awards to facilitate PEER attendance at the SABER annual meeting (there is funding from SABER for six travel awards each year), the mentorship of PEERs by established SABER members, and highlighting PEER leaders in the PEER Spotlights. These actions began to change the culture of the organization, as evidenced by the diversification of the EC and the transition of the PEER Network from a SIG to a committee. The development of inclusive relationships through SIG activities played a crucial role in fostering a culture shift that enabled PEERs to run for and be elected to the SABER EC. The social inoculation framework also supports the change catalyzed by the PEER Network: Highlighting PEERs in SABER and the leadership experience that they obtained in the PEER Network led to greater ethnic and racial diversity in the EC. PEERs in the EC and those leading the SIG served as models of the ingroup status for others from similar backgrounds, which inoculated PEER SABER members from negative stereotypes and

Table 1.2 PEER Network Initiatives and Their Effect on Culture Change

Initiative or Action	*Effect on Diversity, Equity, Inclusion (DEI), and Framework*	
	Social Inoculation Model	*IPF: Societies*
Community-building and networking opportunities	Activities provide an ingroup in the society that inoculates newcomers against negative stereotypes.	Interactions between individuals with similar equity mindsets lead to inclusive relationships.
Activity design based on member priorities	Align activities with the self-concept of historically excluded individuals.	
EER Journal Club	Increase social belonging and self-efficacy through discussions with colleagues from similar backgrounds.	The discussion of DEI-centered journal articles can promote intercultural mindfulness and identity awareness.
PEER mentoring groups		Mentorship and the professional development of those from marginalized backgrounds drives culture change.
PEER Spotlight	Recognition is an influential action that helps to drive culture change.	Evidence of peers in the organization's ingroup inoculates historically excluded members against negative stereotypes.
Safe-space meetings	Increase social belonging and reduce stereotype threat by identification with ingroup colleagues or experts.	Identity awareness and intercultural mindfulness help to develop inclusive relationships.
Transition from special interest group to committee		Influential DEI action that tangibly contributes to changing SABER's culture.
SABER meeting travel awards	Create a critical mass of PEERs to increase social belonging.	Influential actions help to drive culture change.

Sources: Dasgupta (2011) and Leibnitz et al. (2022a).

increasing their self-efficacy. These individuals, termed "DEI change leaders" in IPF: Societies, are an important part of catalyzing cultural change (Leibnitz et al. 2022a).

Recommendations and Discussion Questions

We hope that this case study will help guide others who want to begin changing the overarching culture of their professional society. We believe that the PEER

Network model will apply to other STEM scientific professional societies since these societies face many of the same challenges and opportunities as SABER. We recommend adapting existing infrastructure within the society to fit the needs of the group. In our case, SABER had the existing infrastructure of SIGs, mentoring groups, and travel awards, which were easily adapted for our needs. Further, the familiarity of the existing infrastructure facilitated the coordination of new initiatives. We also recommend maintaining open communication between the new group and the leadership and general membership of the organization. For us, this was key to cultivating a trusting environment around the creation of the PEER Network.

We hope that the following questions will be useful to those who want to adapt our model within their society:

1. Does your organization have an overarching culture that feels exclusive to certain populations of individuals?
2. Does your society have data to determine the potential issues or challenges that contribute to a culture that is not inclusive? If not, societies should plan to begin their work toward equity by collecting these baseline data.
3. Are there individuals within your society who can organize a group to create community among marginalized members? If so, how can these individuals be engaged?
4. How can your society support the individuals leading culture change to promote sustainability and prevent burnout?
5. What is the existing infrastructure within your society that can be leveraged to create the community?
6. What are some of the barriers to cultural change that you anticipate, and how can you proactively address them?
7. How can you secure buy-in from members of the dominant culture in the professional society?

In our case, the social conditions when we created the PEER Network helped with buy-in from a large portion of the SABER membership. Looking forward, we want to determine how to translate the model of the PEER Network to amplify the voices of other marginalized groups in SABER, like two-year college faculty, LGBTQ+ and neurodivergent individuals, and those with disabilities.

Conclusion

We describe the creation of a group focused on the needs and priorities of PEERs within SABER, how this group began catalyzing a culture change in the organization, and the effect of PEER Network on the diversification of SABER's leadership. We analyzed the culture change that SABER experienced using the social inoculation and IPF: Societies frameworks to highlight the key actions that led to change and thus facilitate the adoption of this model by other societies. Finally, we share discussion questions for those who wish to adapt the model described in this case study to their organization.

Note

1 We thank the individuals who had a role in starting and sustaining the work of the PEER Network and who could not be included as authors of this chapter due to publisher limits: Fernando Tenjo-Fernandez, Melinda Owens, Anne McIntosh, Sayali Kukday, Carlos Goller, Sophia Jeong, Madhura Pradhan, Jennifer Garcia, Laurel Lorenz, Stanley Lo, Supriya K, Stacy Ochoa, Kelsie Bernot, Natasha Woods, Christopher Jang, Bianca Salazar, Mary Bonine, Oyenike Olabisi, Patricia Citlaly Jimenez, Sara Brownell, and Natalia Caporale. We also thank Melinda Owens for her help in drafting the initial version of the PEER member survey. We are thankful to Natalia Caporale for her efforts in starting the PEER Network SIG.

Funding Acknowledgment: This work is supported by the National Science Foundation under Grant No. 2334951 to MS, SR and ME G-O.

References

Asai, D. J. 2020. "Race Matters." *Cell* 181 (4): 754–7.

Asgari, S., N. Dasgupta, and N. G. Cote. 2010. "When Does Contact with Successful Ingroup Members Change Self-Stereotypes? A Longitudinal Study Comparing the Effect of Quantity vs. Quality of Contact with Successful Individuals." *Social Psychology* 41: 203–11.

Barabino, G. A., S. T. Fiske, L. A. Scherer, and E. A. Vargas 2023. "Minoritized Individuals in STEMM: Consequences and Responses to Racial Bias and How STEMM Professionals Can Help." In *Advancing Antiracism, Diversity, Equity, and Inclusion in STEMM Organizations: Beyond Broadening Participation*, edited by G.

A. Barabino, S. T. Fiske, L. A. Scherer, and E. A. Vargas, 117–64. National Academies Press. https://www.nationalacademies.org/.

Borman, K., R. Halperin, and W. Tyson, eds. 2010. *Becoming an Engineer in Public Universities: Pathways for Women and Minorities*. Springer.

Dasgupta, N. 2011. "Ingroup Experts and Peers as Social Vaccines Who Inoculate the Self-Concept: The Stereotype Inoculation Model." *Psychological Inquiry* 22 (4): 231–46.

Harte, H. A., J. L. Gilbert, H. H. Chai, H. W. Soled, K. Ofori-Attah, and K. Gunn. 2009. "Perspectives on Facilitating Minority Faculty Success in Higher Education." *Kentucky Journal of Excellence in College Teaching and Learning* 7, article 5: 45–50.

Hulede, I. V. 2018. "Preparing Students for Success in STEM: Role of Professional Societies." *Cell Biology Education–Life Science Education* 17 (3): es14.

Inzlicht, M., and T. Ben-Zeev. 2003. "Do High-Achieving Female Students Underperform in Private? The Implications of Threatening Environments on Intellectual Processing." *Journal of Educational Psychology* 95 (4): 796–805.

Leibnitz, G. M., D. L. Gillian-Daniel, R. Greenler, R. Campbell-Montalvo, H. Metcalf, V. A. Segarra, J. W. Peters, S. Patton, A. Lucy-Putwen, and E. L. Sims. 2022a. "The Inclusive Professional Framework for Societies: Changing Mental Models to Promote Diverse, Equitable, and Inclusive STEM Systems Change." *Frontiers in Sociology* 6: 784399. https://doi.org/10.3389/fsoc.2021.784399.

Li, Y. L., K. Evans, and M. A. Bond. 2023. "Allies as Organizational Change Agents to Promote Equity and Inclusion: A Case Study." *Equality, Diversity and Inclusion: An International Journal* 42 (1): 135–56.

McGee, E. O. 2020a. *Black, Brown, Bruised: How Racialized STEM Education Stifles Innovation*. Harvard Education Press.

NASEM (National Academies of Sciences, Engineering, and Medicine). 2023. *Advancing Antiracism, Diversity, Equity, and Inclusion in STEMM Organizations*. National Academies Press.

NCSES (National Center for Science and Engineering Statistics). 2021. *Women, Minorities, and Persons with Disabilities in Science and Engineering: 2021*. Special Report NSF 21-321. National Science Foundation.

NSB (National Science Board). 2015. *Revisiting the STEM Workforce: A Companion to Science and Engineering Indicators 2014*. NSB-2015-10. National Science Foundation.

NSF (National Science Foundation). 2019. *Women, Minorities, and Persons with Disabilities in Science and Engineering: 2019*. Special Report NSF 19-304. National Science Foundation. https://ncses.nsf.gov/pubs/nsf21321/data-tables.

Offerdahl, E. G., T. Balser, C. Dirks, K. Miller, J. L. Momsen, L. Montplaisir, M. Osgood, K. Sirum, M. P. Wenderoth, B. White, W. B. Wood, M. Withers, and R. Wright. 2011. "Society for the Advancement of Biology Education Research (SABER)." *Cell Biology Education–Life Sciences Education* 10 (1): 11–13.

Page, S. E. 2007. *The Difference: How the Power of Diversity Creates Better Groups, Firms, Schools, and Societies*. Princeton University Press.

Page, S. E. 2017. *The Diversity Bonus: How Great Teams Pay Off in the Knowledge Economy*. Princeton University Press.

PCAST (President's Council of Advisors on Science and Technology). 2012. *Transformation and Opportunity: The Future of the US Research Enterprise*. Executive Office of the President.

Pearson, J., L. A. Giacumo, F. Arvin, and M. Sadegh. 2022. "A Systematic Multiple Studies Review of Low-Income, First-Generation, and Underrepresented, STEM-Degree Support Programs: Emerging Evidence-Based Models and Recommendations." *Education Sciences* 12 (5): 333.

Popo-Olaniyan, O., O. A. Elufioye, F. C. Okonkwo, C. A. Udeh, T. F. Eleogu, and F. O. Olatoye. 2022. "Inclusive Workforce Development in U.S. STEM Fields: A Comprehensive Review." *International Journal of Management and Entrepreneurship Research* 4 (12): 659–74.

Primus, C., A. N. Zimmerman, A. K. Terovolas, K. F. Block, C. G. Brown, M. D. Burton, A. Edwards, C. M. Etson, S. C. Flores, C. Fry, A. N. Guillory, S. L. Ingram, R. McGee, D. L. Neely-Fisher, S. Paxson, L. Phelan, K. Suggs, L. R. Vega, E. Vuong, J. L. Lujan, M. Ramirez-Alvarado, and V. A. Segarra. 2022. "Scientific Societies Fostering Inclusivity in the Life Sciences through Engagement of Undergraduate Scientists." *Frontiers in Education* 7: 757816. https://doi.org/10.3389/feduc.2022.757816.

Segarra, V. A., and C. M. Etson. 2023. "Editorial: Professional and Scientific Societies Impacting Diversity, Equity and Inclusion in STEMM." *Frontiers in Sociology* 8: 1232770.

Segarra, V. A., C. Primus, G. A. Unguez, A. Edwards, C. Etson, S. C Flores, C. Fry, A. N Guillory, S. L Ingram, M. Lawson, R. McGee, S. Paxson, L. Phelan, K. Suggs, L. R. Vega, E. Vuong, J. C. Havran, A. Leon, M. D. Burton, J. L. Lujan, and M. Ramirez-Alvarado. 2020a. "Scientific Societies Fostering Inclusivity through Speaker Diversity in Annual Meeting Programming: A Call to Action." *Molecular Biology of the Cell* 31 (23): 2495–501. https://doi.org/10.1091/mbc.E20-06-0381.

Segura-Totten, M., B. Dewsbury, S. M. Lo, E. G. Bailey, L. Beaster-Jones, R. J. Bills, S. E. Brownell, N. Caporale, R. Dunk, S. L. Eddy, M. E. García-Ojeda, S. M. Gardner, L. E. Green, L. Hartley, C. Harrison, M. Imad, A. M. Janosik, S. Jeong, T. Josek, P. Kadandale, J. Knight, M. E. Ko, S. Kukday, P. Lemons, M. Litster, B. Lom, P. Ludwig, K. K. McDonald, A. C. S. McIntosh, S. Menezes, E. M. Nadile, S. L. Newman, S. D. Ochoa, O. Olabisi, M. T. Owens, R. M. Price, J. W. Reid, N. Ruggeri, C. Sabatier, J. L. Sabel, B. K. Sato, B. L. Smith-Keiling, S. D. Tatapudy, S. J. Theobald, B. Tripp, M. Pradhan, M. J. Venkatesh, M. Wilton, A. M. Warfa, B. N. Wyatt, and S. A. Raut. 2021. "Chronicling the Journey of the Society for the Advancement in Biology Education Research (SABER) in Its Effort to Become Antiracist: From Acknowledgement to Action." *Frontiers in Education* 6. https://doi.org/10.3389/feduc.2021.780401.

Sekaquaptewa, D., and M. Thompson. 2002. "The Differential Effects of Solo Status on Members of High- and Low-Status Groups." *Personality and Social Psychology Bulletin* 28: 694–707.

Siy, J. O., and S. Cheryan. 2013. "When Compliments Fail to Flatter: American Individualism and Responses to Positive Stereotypes." *Journal of Personality and Social Psychology* 104 (1): 87–102.

Skvoretz, J., G. Kersaint, R. Campbell-Montalvo, J. D. Ware, C. A. S. Smith, E. Puccia, J. P. Martin, R. Lee, G. MacDonald, and H. Wao. 2020. "Pursuing an Engineering Major: Social Capital of Women and Underrepresented Minorities." *Studies in Higher Education* 45 (3): 592–607.

Stout, J. G., N. Dasgupta, M. Hunsinger, and M. A. McManus. 2011. "STEMing the Tide: Using Ingroup Experts to Inoculate Women's Self-Concept in Science, Technology, Engineering, and Mathematics (STEM)." *Journal of Personality and Social Psychology* 100 (2): 255–70.

Thiem, K. C., and N. Dasgupta. 2022. "From Precollege to Career: Barriers Facing Historically Marginalized Students and Evidence-Based Solutions." *Social Issues and Policy Review* 16 (1): 212–51.

Thompson, S. K., S. Hebert, S. Berk, R. Brunelli, C. Creech, A. G. Drake, S. Fagbodun, M. E. Garcia-Ojeda, C. Hall, J. Harshman, T. Lamb, R. Robnett, M. Shuster, S. Cotner, and C. J. Ballen. 2020. "A Call for Data-Driven Networks to Address Equity in the Context of Undergraduate Biology." *Cell Biology Education—Life Sciences Education* 19 (4): mr2.

Trapani, J., and K. Hale. 2019. *Higher Education in Science and Engineering: Science and Engineering Indicators 2020*. NSB-2019-7. National Science Foundation.

Womack, V. Y., B. K. Thakore, C. V. Wood, D. C. Jewett, R. F. Jones, S. L. Ingram, J. A. Clark, C. L. Fry, L. Wecker, and R. McGee. 2020. "The ASPET Mentoring Network: Enhancing Diversity and Inclusion through Career Coaching Groups within a Scientific Society." *Cell Biology Education–Life Sciences Education* 19 (3): ar29.

Yen, J. W. 2019. "Advancing Diversity, Equity, and Inclusion Is a Leadership Issue for All of Us." *PLOS Channels and Collections Blog*, November 14. https://collectionsblog.plos.org/advancing-diversity-equity-and-inclusion-is-a-leadership-issue-for-all-of-us/.

Yip, T., C. S. L. Cheah, L. Kiang, and G. C. N. Hall. 2021. "Rendered Invisible: Are Asian Americans a Model or a Marginalized Minority?" *American Psychologist* 76 (4): 575–81.

Zambrana, R. E., R. Ray, M. M. Espino, C. Castro, B. Douthirt Cohen, and J. Eliason. 2015. "'Don't Leave Us Behind': The Importance of Mentoring for Underrepresented Minority Faculty." *American Educational Research Journal* 52 (1): 40–72.

2

Enhancing Professionalization in Computing: A Case Study of Staff and Faculty Practices at Hispanic-serving Institutions

Nancy Amador, Raena Cota, Nohelia Diplan,
Sarah Hug, and Melanie Martin

Participation in professional societies and conferences is vital for undergraduate students as they develop their scientific professional identities, yet these opportunities are out of reach for many students at Hispanic-serving institutions (HSIs) in the United States.[1, 2] As the demographics in the United States' college-going population continues to shift, the number of HSIs grows annually, yet research in higher education indicates a difference between *enrolling* Hispanic students in great numbers and *serving* the Hispanic/Latine population. The designation, defined by 25 percent Hispanic enrollment at a higher education institution, is consequential because federal funding can be accessed for institutions with the label "HSI." Although traditionally underfunded and overlooked in federal, state, and regional policy efforts, HSIs play a central role in shaping the ecosystem of the nation's STEM workforce by preparing disproportionately high shares of Latine, African American, and Native American STEM graduates (Espinosa et al. 2019), many of whom are first in their family to earn four-year degrees. While only 11 percent of all bachelor's degrees awarded in computer science (CS) nationally are granted to Latine students, at HSIs 37 percent of the bachelor's degrees in CS are awarded to Latine students (Espinosa et al. 2019).

In this case study, we describe how national and local practices support student success at a professional conference. We use the concept of servingness to frame the approach that Computing Alliance of Hispanic-Serving Institutions (CAHSI; https://cahsi.utep.edu/) staff and faculty take toward supporting computing students to apply for, attend, and present work at a professional conference. In this chapter, we build on the work of Kezar and Holcombe (2017) to structure the descriptions of the specific actions provided by staff and faculty to prepare and guide postsecondary student success in their professional conference experiences.

Students with limited funds are less likely to be able to afford attending professional conferences, and students first in their family to attend college may not know why professional conferences are important for their careers. Because students at HSIs are more likely to work and have family care obligations, their out-of-class time is limited—they may be less likely to conduct research with faculty, which is a typical entrée into professional society participation. Not only are funds scarce for travel to professional conferences, but first-generation students may also have fewer travel experiences and need guidance regarding airline travel and hotel stays. In this multisite case study, we address the following two questions:

1. How does a national alliance create a structure for HSI student success at a national conference?
2. What postsecondary institution staff and faculty practices can support successful undergraduate engagement in professional CS conferences?

Aaron Mendez[3] did not complete high school and dropped out in his second year; he didn't think of school as a safe place, nor did he do well in passive classroom settings. After taking some time off, he joined his friend in some technical pursuits. They worked on 3D printing projects together. Eventually Aaron enrolled in a four-year CS program at a local university.

For the authors, stories like Aaron's are not uncommon. Students who attend open-access regional universities arrive with varied educational, technical, and professional experiences. The extent to which professional conferences open doors for Latine[4] students is vast—and the ways in which professional conferences expand a family's access to cultural, financial, and social capital cannot be overstated. Creating departmental cultures of access to high-level, CS conferences shifts upward mobility. For students attending under-resourced collegiate HSIs, positive conference experiences expand their understanding of what is possible in STEM.

Background of the Study in a National Alliance Context

Structures Designed to Support Student Conference Participation

The CAHSI is an NSF National Inclusion across the Nation of Communities of Learners of Underrepresented Discoverers in Engineering and Science (INCLUDES) and Broadening Participation in Computing (BPC) Alliance. CAHSI's mission is to "Grow and sustain a Networked Community committed to recruiting, retaining, and accelerating the progress of Hispanics in computing." CAHSI's vision statement is, "By 2030, Hispanics will represent 20% or more of those who earn credentials in computing." In this case, credentials are defined as degrees and certifications that lead to gainful employment and advancement in the field. CAHSI is an alliance of postsecondary HSIs in the United States and Puerto Rico. Members of the alliance are collegiate, though the alliance itself serves as a unified voice that partners with industry, nonprofits, K–12 education, and policymakers to promote equity in CS. CAHSI has centralized staff within the organizational headquarters as well as regional staff and leadership in the North, Southwest, West, and Southeast, where the United States has the highest concentration of Hispanic students, and thus HSIs. The four CAHSI regions are led by New Mexico State University (NMSU) in the Southwest (where Cota serves as CAHSI connector); Kean University in the North (where Diplan is CAHSI coordinator and Amador was previously CAHSI connector); University of Puerto Rico, Mayaguez for the Southeast region; and California State University, Dominguez Hills for the West region. All college and university members sign a memorandum of understanding (MOU) and designate a CAHSI faculty lead (such as Martin, who is associate faculty at California State University, Stanislaus [CSU Stanislaus]). Strategic actions related to CAHSI goals occur at the national, regional, and local (or departmental) levels.

Great Minds in STEM (GMiS) is an independent organization that hosts the GMiS Conference, which promotes STEM career development and recognizes minority contributions in STEM. CAHSI co-locates the CAHSI Summit at the GMiS Conference, partnering with GMiS to support its mission. GMiS engages a great deal with STEM professionals who are Hispanic, while CAHSI membership is primarily collegiate. Through its involvement in the GMiS Conference, CAHSI amplifies its outreach, enabling its members to access networking, mentorship, and scholarship opportunities specific to Hispanic students pursuing STEM fields. Although the GMiS Conference is not directly run by CAHSI, the alliance actively collaborates in many capacities to enrich the experiences of students and professionals alike.

As CAHSI became an INCLUDES Alliance, it shifted strategy toward collective impact (Hanleybrown et al. 2012). Guided by this approach to change management, the CAHSI INCLUDES Alliance looks for the following conditions when considering partnerships with other organizations:

- *Common agenda*—A common understanding of the problem and a shared vision for change.
- *Shared measurement*—Collecting data and measuring results with a focus on performance management and shared accountability.
- *Mutually reinforcing activities*—Different approaches with coordination through a joint plan of action.
- *Continuous communication*—Consistent and open communication with a focus on building trust.
- *Backbone support*—Separate organizations with staff, resources, and skills to convene and coordinate participating organizations.

In the following sections, we describe the ways that CAHSI expanded capacity to promote student participation in a professional conference and how a CAHSI partnership with a professional conference organization with similar goals and values collaborated to enhance professional opportunities for students (Figure 2.1). We also detail how the alliance leveraged grant funds to scaffold student success in the conference and their subsequent academic and career pursuits.

Making the Conference Accessible to More Students—Leveraging CAHSI National Activities to Support GMiS Participation

Two additional CAHSI student programs link student participation with the GMiS Conference: CAHSI Student Scholars Leadership Program and CAHSI Advocates Program. CAHSI Student Scholars are CAHSI students who have shown outstanding achievements in the classroom and in the community. Selected scholars work with the CAHSI Backbone (the leadership group described in the collective impact model) and their local CAHSI representative to develop a work plan and coordinate efforts around student success while engaging their peers. Selected scholars are announced at the annual GMiS Conference. CAHSI student advocates have a leadership

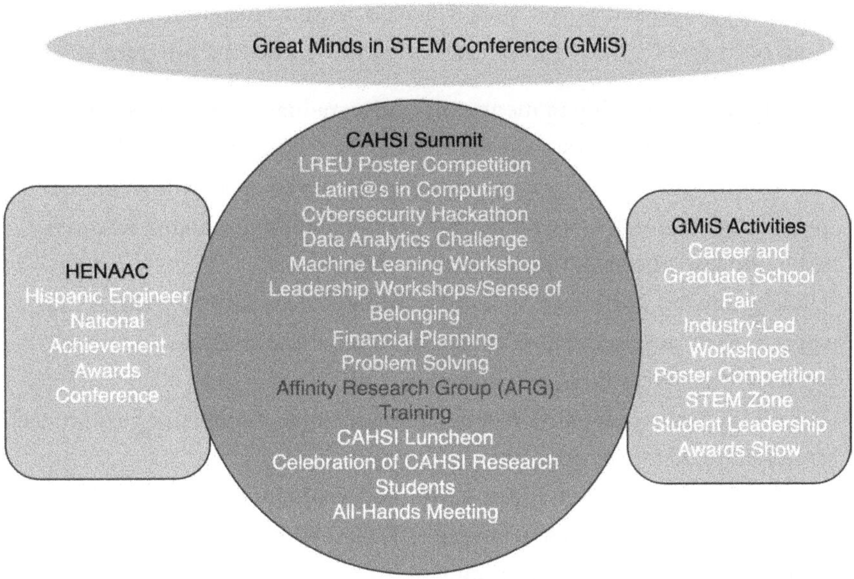

Figure 2.1 Collaboration across organizations to support student conference success. Created in PowerPoint by the authors of this chapter.

role, attend GMiS annually, and receive specialized training in community building, student engagement, and messaging about student opportunities. The CAHSI Advocates Program engages student leaders to serve as advocates in support of student success through promotive interactions and the sharing of CAHSI practices and activities. The program creates a virtual network of student leaders that fosters community building. CAHSI provides at least one workshop at GMiS specifically for the CAHSI scholars for training and community building.

National Program Scaffolds Conference Participation through a Local Research Experience for Undergraduates for Students from Historically Excluded Groups in Computing

CAHSI has designed and implemented a structured co-curricular pathway toward GMiS participation and career development through the Local Research Experience for Undergraduates (REU) program, funded by a grant from the NSF BPC Alliance Expansion. The Local REU program is supported by CAHSI regional staff and faculty, including the authors of this chapter. The national program supports faculty and staff efforts to recruit Hispanic undergraduates with no prior research experience into a structured, national program in which:

- Faculty receive training in mentoring undergraduate researchers that is modeled after CAHSI's signature practice, the affinity research group model (Gates et al. 1999). Mentors are encouraged to develop a growth mindset perspective and lack of assumptions regarding undergraduates' research skills and knowledge as they enter the program.
- Undergraduates participate in local workshops and national webinars to support their development of career knowledge and the graduate school options available to them in computing.
- The GMiS poster competition application serves as a collective goal for each undergraduate researcher.
- The common expectation is that each undergraduate researcher will present and attend the professional conference, GMiS, where they receive extended professional development training.

The Local REU has diversified and increased student participation in the GMiS Conference by expanding representation from more CAHSI institutions. As Local REU student opportunities involve multiple students per campus, the program provides a critical mass of students from historically excluded groups who can get to know one another before they attend the conference in the fall, which builds a sense of community among peers in their computing departments. In addition, the Local REU repeating students are positioned as near-peer mentors who can support application to the Local REU program as well as the conference; near-peer mentors can serve roles in student leadership and help new conference attendees prepare for traveling to a professional conference. Through the structured Local REU programming, students are required to submit a poster for the GMiS student poster competition, thus improving the number of submissions for this GMiS activity.

Since the inception of the Local REU program, Local REU students have become a large portion of the poster competitors at GMiS, estimated at nearly half of the undergraduate competitors in 2023. Winners included community college students, who are traditionally removed from undergraduate research opportunities in CS due to the lack of faculty research mentors available at two-year institutions. The CAHSI Local REU program allows for deeper exploration of the research process by providing financial support for students to work on research projects during the school year. Additionally, CAHSI enables more sustained mentorship by pairing CAHSI students and faculty mentors from the same institution. However, students also have opportunities to participate in a virtual Local REU project with faculty from other CAHSI institutions if the student's research interest lies outside the faculty at their own institution.

Creating Cohorts of Engaged CS Students—Leveraging S-STEM Programs

Over the past decade, CAHSI computing departments have supported one another in the development of scholarship alliance grant proposals to the NSF Scholarships in STEM (S-STEM) program. To date, collaborators have been successful in earning three collaborative S-STEM grants that serve three of the four regions of the CAHSI organization (North, Southwest, and West). A major activity of each funded S-STEM grant includes conference participation at GMiS—the S-STEM grant sets aside money so that each scholar attends the conference at least once. This element of the S-STEM programs supports local community building, assures that S-STEM students receive high-quality professional development, and creates a greater motivation for students to apply to the scholarship. S-STEM evaluations and social science research continuously find that the GMiS Conference was an important aspect of students' computing career development and that traveling with their peers from their college cemented peer relationships that were valuable to their CS success (Hug 2018, 2023; Hug et al. 2024).

High-quality Professional Organization Partnerships

CAHSI formed a partnership with GMiS[5] to co-locate the professional conference serving students and professionals in engineering and CS with the CAHSI Alliance annual meeting, called the CAHSI Summit. This connection was forged through social and professional networks of practitioners serving Hispanic students (e.g., the GMiS organization was honored with an award from a prestigious national policy organization[6] the same year as CAHSI). The first year of widespread participation at the GMiS Conference, 124 students from CAHSI institutions attended—at the 2023 conference, 351 CAHSI students attended, or 283 percent of the 2016 attendance number. In addition to student funding for conference attendance, CAHSI prioritized staff and faculty attendance at GMiS—approximately 60 CAHSI-affiliated faculty and staff members attended in 2023.

Over the years, the CAHSI alliance has connected with multiple conference organizations—other collaborations have waned and the GMiS partnership strengthened. The characteristics of the CAHSI–GMiS partnership are detailed below. These characteristics contributed to the partnership's longevity.[7]

- *Shared core values and equity understanding*—The two organizations share core values and understand issues of equity in STEM similarly. Both organizations focus on the career development of Hispanic STEM students.
- *Student recruitment and support*—The co-location of GMiS and CAHSI helps to recruit student participants for the conference and provides some funding for the travel and registration of students via GMiS scholarships as well as CAHSI funding for students who attend CAHSI institutions. In turn, CAHSI boosts the representation of CS and computer engineering student attendees at GMiS. CAHSI also develops a system of support to produce student research posters for the conference, which boosts the numbers of student poster competitors at GMiS.
- *Transparent fundraising opportunities*—Both organizations have independent opportunities to fundraise with clear transparency regarding which funds are received by each organization.
- *Defined professional development roles*—Professional development content responsibilities are clear and well defined. Opportunities are expanded for CAHSI students through CAHSI-driven, computing-specific, professional development sessions.
- *Flexibility and continuous communication*—CAHSI is included in GMiS leadership meetings that involve planning the content and the calendar of events. This engagement helps the GMiS Conference grow alongside CAHSI's participation as a co-located meeting.
- *Asset-based thinking as a common approach*—Faculty, staff, and career fair-personnel exhibit asset-based thinking about students and create opportunities for growth in all interactions with students.

The collaboration has been successful, though it has had its drawbacks. Attendance capacity is an issue for many of the CAHSI technical events such as an annual hackathon, meaning not all of the students who want to attend specific technical workshops are able to do so. Overlapping student tracks at conferences means that students cannot take advantage of all of the offerings and must make decisions related to their professional goals. These decisions are often navigated with guidance from staff and faculty at their postsecondary institutions.

CAHSI Summit–GMiS Conference Activities Developed to Explore Computing

CAHSI hosts three workshops and competitions for students at GMiS: the Cybersecurity Hackathon, Data Analytics Challenge, and Machine Learning Workshop. All CAHSI students attending GMiS are expected to participate in one of these events. Most students participate in the Cybersecurity Hackathon, which consists of a series of workshops that allow novice participation followed by a team competition. The event serves many purposes: introducing students to cybersecurity in a meaningful way, providing a mentoring environment with other students, and team and cohort building. The data analytics and machine learning workshops were added in recent years to provide experience with upper-level topics in an accessible way to students earlier on in their CS careers. Each has an element of competition and hands-on exploration.

The CAHSI Latin@s workshop includes a virtual pre-conference workshop with a panel discussion on networking and the hands-on creation of a networking plan for the conference. Then, an in-person workshop at the conference provides an opportunity to network with industry representatives and discuss community building for career success. Attendees implement their networking plan at the conference and write a structured reflection on it after the conference.

Case Review of Relevant Literature: Servingness for Intersectional Populations—First-generation, Latine College Students

Aaron started doing well in his CS courses. In his first year, he saw a flier that offered $200 for students willing to try CS research. He applied and got a position.

Research in higher education has begun to identify an institutional characteristic of HSIs that support the success of Latine students in higher education, described as “servingness” (Garcia et al. 2019; Núñez et al. 2015). Servingness is a concept that focuses on the capacity of an organization to create equitable opportunities for Latine students to complete their degrees at similar rates as students from dominant groups. In addition, servingness entails engaging the Latine community through culturally relevant and responsive pedagogy (Garcia et al. 2019). *Latine serving* organizations build on the cultural wealth of Latine student populations to produce outcomes at similar rates to other student demographic groups. Using the cultural practices of Latine students as a resource for deepening rather than diminishing connections to professional practice, serving organizations foster students' academic and non-academic success (Gutiérrez 2008; Núñez and Gildersleeve 2016). Identifying serving practices within professional organizations is a concrete way to increase access and define what is meant by professional growth in a more expansive and inclusive way.

Creating equitable outcomes for all students in STEM requires additional support systems, particularly for students who are the first in their family to earn college degrees.[8] In a study of staff and faculty in the CAHSI network, Rivera and Núñez (2022) found that it was vital to rework practices from the predominantly white institution (PWI) models to effectively support a different student population. According to the authors, “rather than imitating what highly selective, predominantly white institutions (PWIs) do, CAHSI departmental personnel develop approaches that are grounded in their own institutional realities and those of their students, which often do not resemble those of PWIs” (Rivera and Núñez 2022, 39). In fact, the needs of first-generation college students studying STEM in higher education are varied, and span developmental, financial, emotional, academic, and advisory realms —the bulleted list below is adapted from the work of Kezar and Holcombe (2017) regarding the complex needs of first-generation STEM students at an HSI. First-generation college students benefit from explicit expectations for student behavior, access to structured academic pathways, intrusive advising that is proactive rather than passive and reactive, explicit instruction regarding how students traverse academic majors toward professional opportunities, and assignment or structured engagements with peer mentors in the major. CAHSI stakeholders envision the GMiS Conference as a vehicle for supporting student career development via access to professional networks and job opportunities beyond their postsecondary institution and have imbued servingness practices into the preparation and guidance surrounding the conference. The CAHSI staff and faculty who provide guidance to prepare students before, during, and after the professional conference incorporated many of the following elements of serving into their work with students attending GMiS.

First generation-specific needs

- Family support
- Structured academic pathways
- Working with faculty
- Setting expectations
- Creating college knowledge
- Intrusive advising

Computer science-specific needs

- CS advising
- Hands-on, experiential learning
- CS career awareness
- Peer mentors in the major
- Develop a CS identity
- Peer-based learning models

To effectively support first-generation college students in STEM, Kezar and Holcombe (2017) highlight several key practices. Working with faculty is crucial, as faculty can provide mentorship and model professional behavior. Setting clear expectations for student behavior helps students understand what is required for success. Intrusive advising, which is proactive rather than reactive, ensures that students receive guidance before they encounter difficulties. Structured academic pathways help students navigate their courses and stay on track for graduation. Peer-based learning opportunities allow students to collaborate and learn from one another. Developing a strong science identity is facilitated through experiences that enable students to see themselves as capable scientists, such as performing well in computing and receiving recognition from peers and professionals. Peer mentors offer additional support, sharing their experiences and advice. Hands-on and experiential learning, including internships and research opportunities, provides practical experience and reinforces classroom learning. Finally, increasing STEM career awareness helps students understand the range of opportunities available to them and how to pursue these careers. The CAHSI staff and faculty integrate many of these elements into their preparation of students for professional conferences like GMiS, which fosters a comprehensive support system that addresses the multifaceted needs of first-generation STEM students. We use this framework to highlight the support structures that CAHSI develops within institutions to support positive conference experiences for HSI students.

Aaron began a second research experience as a sophomore. He had never left the state, yet he was offered an opportunity to go to a conference in Washington, DC with his research mentor. His first-semester instructor began to ask Aaron to speak to her new students about what it was like to do research as a first-year student. He regularly shared about his lack of confidence but eventual success in research projects early in the degree program.

Findings

Professional conferences serve multiple developmental purposes for students with differing needs and at different points in their careers. For freshmen and sophomores, a professional conference can give them a chance to understand potential career subfields in computing, such as cybersecurity or artificial intelligence, and a discussion with industry professionals may serve to guide future course selection or research endeavors. For a student feeling "othered" in the department, a conference experience with peers can help create a social network back at home that will support future academic success. For students with undergraduate research experience, a conference can provide a connection to a graduate school opportunity or a job prospect. When considering conference student impact, the authors draw on multiple academic, social, and professional goals that students may have along their computing career pathway and measure many indicators of impact to capture all potential conference goals.

Case study results appear below with quantitative survey data that initially indicates the impact of attending the conference and qualitative, emergent thematic data that indicates the effective strategies employed to support students from the authors' institutional contexts.

Quantitative Results

CAHSI–GMiS student experiences have been studied by the external evaluation team[9] since CAHSI began its partnership with the organization in 2016. Over time, student post-survey evaluation results have remained consistent, with a slight dip in effectiveness during the online implementation of the conference during the height of the Covid-19 pandemic. GMiS survey results from fall 2022 (when the conference returned to in-person) are summarized in this chapter in brief, with eighty-four respondents providing feedback.[10] The survey is sent to registrants four to eight weeks after the conference experience. Because of this planned delay, we are able to learn about any activities students engage in *following* the conference. Questions in the evaluative survey relate to (a) preparation for the conference by CAHSI staff and faculty before attendance, (b) CAHSI-led workshops at the conference and the approach to learning through those workshops, (c) social impact of conference attendance, and (d) career development impact of attending the conference. Only conference attendees who received CAHSI or CAHSI-related funding to attend the conference received the CAHSI evaluative survey each year.

Servingness Approaches as Evidenced in CAHSI–GMiS Programming

CAHSI and GMiS have a shared vision of students from HSIs as having great potential for excellence and believe in the need for equity and inclusion to occur together in computing education and practice. This vision was deliberately integrated into the development and facilitation of CS technical content. A six-item, six-level, Likert survey scale related to these concepts was developed to understand whether and how facilitators maintained these values and pedagogical practices. The items are as follows:

1. The sessions provided adequate preparation.
2. The sessions created opportunity to build community.
3. The session leaders emphasized growth and improvement more than previous knowledge.
4. The session leaders gave us ways to actively participate throughout the sessions.
5. I developed technical skill/knowledge in the technical sessions.
6. I developed confidence in my abilities during the technical sessions.

In the 2022 CAHSI evaluation results for GMiS impact, 86 percent or more of participants responded positively (agreed a little, agreed, or agreed strongly) to all of the statements regarding servingness values and approaches to student development. The highest rated items were items 1 and 2, with 92 percent agreement for both.

Overall Impact of the Conference—Recent Results from Participant Surveys

Nearly all survey respondents stated that the conference at least "somewhat" enhanced their career development (99 percent).[11] Participants said the event "greatly enhanced" their career development (51 percent), while 33 percent said it supported their career development and 15 percent said GMiS supported their career development "somewhat." Responses regarding the GMiS Conference and its ability to support student progress indicated that all or nearly all of the eighty-four student survey participants experienced: increased interest in an industry career (100 percent), increased knowledge of the job application and interview process (99 percent), increased dedication to the computing major (99 percent), help feeling more connected to peers in the CAHSI community (100 percent), increased knowledge of career pathways in the field (99 percent), more opportunities to get career advice (100 percent), and increased interest in a particular area of computing (96 percent).[12]

Nearly all attendees met new peers (99 percent), most met new faculty members (81 percent), and 71 percent applied to a technical position while at GMiS. Attendees (59 percent) spoke with at least one job fair recruiter at the GMiS Conference in an interview. Graduate school representative interactions were less frequent (43 percent). This may relate to the greater representation of industry at the conference overall. Students received internship offers at the conference (32 percent) and 20 percent received a job offer at the conference.

As a junior, Aaron attended the GMiS Conference and presented his work in the student poster presentation. He won the competition. He interviewed for an internship at one of the networking events—he mentioned that he kept joining the line for one-on-ones so he could maximize the number of conversations he had with professionals. He secured an internship with a major technical firm for the following summer.

CAHSI Support Led to Greater Preparation for the Professional Conference

Students were prepared locally through communal activity as well as individualistic activity before attending GMiS. Of those who received or took on some kind of preparation: most students worked on their résumés before attending GMiS (84 percent), while about half developed a goal for the conference (51 percent), half asked permission to miss class to attend the conference online (51 percent), and slightly fewer asked permission to miss work to attend GMiS virtually (42 percent). Others had access to CAHSI-related content and local networks to prepare for GMiS—this included meeting with CAHSI club members prior to attendance (42 percent), attending faculty-led webinars or meetings to prepare for the conference (49 percent), or attending CAHSI student-led webinars or meetings about the conference (43 percent).

The preparations, like those detailed in this chapter, appear to influence outcomes following the conference—specifically, those who received preparation from peers had higher average scores on outcome items. For the items shown below, differences between peer-led preparation and those without peer-led preparation are statistically significant with moderate effect sizes; in all cases, peer-supported student averages are higher.

- Increased my interest in an industry career ($p = 0.009$; effect size Cohen's $d = 0.641$).
- Increased my computing skills ($p = 0.017$; effect size Cohen's $d = 0.591$).
- Increased my knowledge of career pathways in my field ($p = 0.008$; effect size Cohen's $d = 0.663$).
- Increased my interest in graduate school programs ($p = 0.021$; effect size Cohen's $d = 0.573$).

Qualitative Results

The author team built on shared experience as CAHSI staff and stakeholders to discuss the practices used to support student success over time in the alliance. Authors have longevity with CAHSI ranging from 5 to 17 years. Author participants wrote out their institutional perspective on supporting students to attend and thrive at conferences, with multiple online meetings held to make collaborative meaning. Following two, focused, online meetings regarding GMiS support, the Kezar and Holcombe (2017) framework was selected as a way to structure the themes that were emerging from the discussions. The qualitative results reflect collaborative meaning-making and highlight similarities and differences across sites as applicable.

Communicating Expectations Explicitly

All three institutions highlighted in this case study communicate explicit expectations for students, faculty, and staff for the GMiS Conference to maximize positive and successful experiences at the conference. These expectations span pre-conference, conference, and post-conference timeframes. While the implementations may differ by institution, there is a strong commonality of overarching goals among the three institutions. These common goals include establishing communication channels, providing a framework for what students can expect in their professional conference travel, and teaching and supporting soft skills to maximize positive conference experiences.

Before pre-conference expectation setting can begin, each institution has a process to select students and secure their commitment to attend the conference. The recruitment process varies in part due to student constituency and funding sources. At each institution there are CAHSI students engaged in local leadership or research and who are expected to attend GMiS—for example, the CAHSI Student Advocate, CAHSI Student Leadership Scholars, CAHSI Local REU participants, a subset of CAHSI Student Club members, and a subset of NSF S-STEM Scholars. Each institution may have other students or student groups that they tend to tap for recruitment (e.g., departmental cybersecurity club, previous CAHSI scholars). In all cases, this is largely through faculty and staff recommendations. Some examples of other students are student leaders not directly affiliated with CAHSI, students participating in CS faculty research groups, or students who faculty believe could benefit significantly from the experience.

Intentions to build conference participation from a variety of experience levels is evident in registration data—this multi-year recruitment helps support peer mentorship, described later in the case study. Staff and faculty have found that GMiS attendance for freshmen can be highly motivating for them to plan a college roadmap that includes undergraduate research, internships, and other projects on a path to graduate school or a competitive CS job. Sophomore and junior students often get summer internships with companies participating in the career fair. The authors have also found that well-prepared seniors may leave the conference with multiple job offers.

Once students are selected, usually two to three months before the conference, communication is established, and preparation begins. Within the framework of the common goals, each institution leverages its existing structures for serving students (Hug et al. 2021) and includes more as needed.

For example, at CSU Stanislaus, S-STEM scholar meetings are expanded to include conference participants and build up to the conference in late fall; at Kean, additional webinars and workshops are added to the calendar and communicated visually, in person, and in class. Each institution has used communication software (Discord, Teams [NMSU], GroupMe, Canvas [CSU Stanislaus], Whatsapp, and Discord [Kean]). Authors found it helpful to get all participants ready to use these platforms, not only to maximize communication leading up to the conference but so everyone is comfortable with using them during the conference.

All conference attendees need to complete local university travel paperwork and often additional CAHSI paperwork. At CSU Stanislaus we have found it most effective to complete this paperwork as a hands-on group activity with faculty and staff available to answer questions and review documents. Staff and faculty that support student conference participation do not assume that students have travel experience, particularly without family present, so we provide a combination of verbal and written instruction from faculty, staff, and more experienced student mentors with sufficient detail and time for questions. Some examples of these topics are: how to dress, how to travel (including plane or train, luggage, and documentation), how to stay in a hotel, how to find their way around Pasadena (conference host city, including restaurants, grocery and drug stores, and thrift stores for emergency clothing purchases), and how to present themselves in new public and professional settings.

In addition, there are pre-conference training topics that are more specific to making the most of a professional conference, including writing a good resume, how to network, how to interview, and how to navigate the conference program and choose events. Authors across institutions agree that preparation is most effective when it involves other students, live demonstrations, and hands-on activities.

Upon arriving at the conference travel-weary students need to register, check in to hotels, locate morning venues, and find food. CSU Stanislaus faculty shepherd students through this process to ensure that everyone is settled and then invite all the students to a hosted dinner at a local restaurant. CSU Stanislaus faculty have found the dinner to be a cohort building event that allows faculty to see the forming social groups, ensure no one is being left out, and serves as a good opportunity for reminders about upcoming events.

During the conference all three institutions work to provide logistic support, problem-solving, as well as social and emotional support for students. We work to anticipate student needs and potential pitfalls at the conference. Kean staff bring water bottles and snacks to conference venues. NMSU and CSU Stanislaus organizers print research posters for students who were not aware they needed to print them before traveling to the conference. Kean staff and CSU Stanislaus faculty rove the conference venues, while NMSU staff inform students where to find them throughout the day. All of the attendees have met the chaperone staff and faculty on campus, and we find that chance encounters with students provide opportunities to provide support, encouragement, and congratulations when things are going well.

CAHSI provides a group photo session where everyone wears a CAHSI t-shirt, and each CAHSI school also takes a school-only group photo. In addition to cohort building, this event provides a check-in opportunity midway through the conference. In the background, faculty and staff monitor electronic communication platforms to provide reminders, clarification, and advice as needed. Kean has morning checks, group meetings (pre-departure), selfie check-ins,[13] and group photo check-ins. CSU Stanislaus faculty monitor the student communication to make sure that everyone follows the buddy system (no one left out or left behind) and that all students are where they need to be. NMSU staff provide personal contact information so conference attendees can reach out with questions, concerns, and celebratory items, such as getting a job or internship interview before the job fair occurs.

After the conference all sites described the importance of student reflection on what went well and what could be improved. All institutions solicit student feedback and reflection. All also provide opportunities for students to be recognized for their accomplishments and encourage other students to attend in the future. Kean, for example, has a storytelling panel event and creates a poster that recognizes student achievements and awards at the conference. CSU Stanislaus continues to meet weekly with the S-STEM students after the conference, which reinforces the community building that takes place at GMiS. NMSU checks in personally with students who attended to understand how they fared in securing internship, graduate school, or job interviews and offers.

As a senior, Aaron continued sharing his experiences with his peers and near-peers in the CS program. As a senior he won Researcher of the Year at his university. His internship had paved the way for a full-time technical position at a major firm. He credits the support from his department and his research mentors for getting him through the program and onto a strong career.

Feeling Like Family—Building a Sense of Community at the Conference

At CAHSI, the feeling of *familia* is cultivated throughout. A common sight at CAHSI events is one of staff, faculty, and students embracing, asking after children, and sharing a meal or a beverage with one another outside of the formal itinerary. The CAHSI formal agenda also enhances the sense of community at GMiS. This occurs through the design of sessions specifically tailored to build student confidence and bring together cohorts of CAHSI students. For example, sessions provide opportunities to practice whiteboarding ahead of technical interviews guided by more knowledgeable others, and are structured to focus on subsets of students, such as Latinas in Computing. Student leaders are encouraged to share with one another how they work locally to create a sense of familia or *familismo*. Familial support in academic life focuses on systems of encouragement, guidance, and assistance that individuals receive from their family members and close-knit community, which significantly impacts their educational journey and professional development. This support often includes emotional backing, financial assistance, mentorship, and the fostering of a positive academic environment. In the context of HSIs, familial support extends to the institution's culture, where a sense of community and interconnectedness mirrors the traditional family structure. This support system helps students, faculty, and staff to navigate academic challenges, celebrate achievements, and maintain a balanced life, which ultimately contributes to their overall success and well-being within the academic setting.

Creating a sense of familial community in HSIs involves fostering an environment where students, faculty, and staff feel a deep sense of belonging, mutual respect, and support. This sense of community can be cultivated through various practices in supporting professional conference engagement. For example, creating physical and virtual spaces where members of the institution can collaborate, share ideas, and engage in meaningful discussions fosters a sense of community and togetherness. At the conference, groups developed patterns of meeting in specific spaces (e.g., the CAHSI hospitality room, the lobby near the CAHSI workshops). Involving students, faculty, and staff in decision-making processes ensures that everyone's voice is heard and valued to reinforce a sense of belonging and collective responsibility.

Staff and faculty in some cases took on familial roles with students, displaying an ethic of care (Noddings 1984) at the conference. In contrast to faculty and staff who participate in parallel with students, those who took on a familial role anticipated students' needs and identified potential barriers to success. This was done by providing students water and snacks at their poster sessions; motivational support via pep talks to students before presentations; coordinated group outings (such as dinners or lunches); reminding students to take care of their physical, mental, and emotional well-being by encouraging breaks; and problem-solving any last-minute logistic issues that arise where student can become overwhelmed. As an example, staff and faculty worked with students to realize last-minute poster printing when conference presentation expectations were less clear to students. By implementing practices that support community aspects of professional conference attendance, staff and faculty at HSIs develop a familial community that supports the personal and professional development of its members and fosters a nurturing and inclusive academic environment.

Peer Mentoring

As CAHSI institutions foster a culture of conference attendance locally, multiple cohorts of undergraduate GMiS attendees are cultivated. Crafting informal and formal roles for the returning students as peer mentors can be beneficial to undergraduates attending conferences. The staff and faculty who send students collectively have historical knowledge regarding which students attended in past years—they leverage that knowledge to help make connections among students. Some staff build roommate pairs based on whether students have attended in the past—those who have attended before take on the responsibility to support and encourage their near-peers with less experience attending professional conferences. Those near-peer mentors also are asked to work on recruitment by sharing their experience in club meetings, workshops, and webinars—describing their initial concerns and how they overcame them is vital to the inclusion of students from similar circumstances. Students who speak with their peers are encouraged to be vulnerable and describe what made them nervous to apply or what obstacles they faced, such as explaining the need to attend the conference to family members. Staff and faculty model the encouragement and support for less experienced students, with the expectation that those funded in years past will likewise view their role as one of mentoring and guidance for new attendees.

Many students, especially those who have never attended a professional conference before, may be hesitant to attend non-CAHSI, GMiS-sponsored events within the conference, such as the academic bowl or industry-led networking events. However, by sharing information they can form supportive groups across conference events, even during interviewing sessions. Previous GMiS attendees often wait in the interviewing line to provide encouragement to students who may be interviewing for the first time. Having a companion wait with them makes students feel more comfortable and better prepared, as they can practice their elevator pitches while waiting. This practice becomes standard, as those who received the help from peers choose to give back the support to newer attendees.

Discussion

Students served by HSIs, who are often first in their families to attend college, need structures of support to identify appropriate professional venues, apply for funding to attend, and traverse the bureaucracy of institutional travel funding to successfully participate in professional conferences. Assumptions that staff and faculty may have regarding student access to travel knowledge, financial processes, research expectations, and professional pathways can stall the progress of those who may most benefit from professional conference participation. By making all expectations explicit and providing contextual knowledge about professional conference attendance, staff and faculty may better support equity in conference participation. This student support takes time, and in some settings is conducted as invisible labor, or unpaid work, by dedicated faculty and staff who are also non-traditional in STEM in some way, as women of color, first-generation college students, or part of the LGBTQ+ community (Hamblin et al. 2020). Through CAHSI and some departmental and institutional efforts, funds have been established to administratively support staff and faculty in these equity efforts. Local practices must be supported to endure and make a difference in who participates in the professional computing community.

Through the GMiS–CAHSI partnership, staff and faculty are encouraged to attend to what Kezar and Holcombe (2017) call the needs of first-generation STEM students. In the contexts of CAHSI institutions in this case study, first-generation and Latine are overlapping identities that many students attending

GMiS share. The case study indicates that staff and faculty attend to student needs by setting clear expectations for the conference, providing a feeling of *familismo* so all students are accountable to one another and supportive of one another's goals, and build on continuing students to provide peer mentoring opportunities before and after the conference. Quantitative survey results indicate that the conference is impactful for student attendees and that students who are prepared to attend the professional conference by their peers achieve greater gains; providing avenues for students to support peers in their professional conference endeavors may be fruitful for improving outcomes.

Conclusion and Reflection Questions

Guiding students toward successful conference participation takes time and energy, often considered invisible labor that is inequitably distributed in academic settings and often falls on the backs of those who are othered in the professional STEM fields. CAHSI's staffed positions, distributed across institutions that lead the CAHSI regions, were created in part to ease the burden of faculty workloads in supporting students in conferences. In addition, student support grants are leveraged in CAHSI sites to fund faculty efforts in gathering students to prepare for the conference (e.g., S-STEM faculty coordination salary). Reflection questions for the reader include:

1. How can we leverage grant funding to better support diverse students to attend professional conferences?
2. How must academic value systems adjust toward student equity? What could that look like?
3. What systemic efforts would support students in their professional conference success? How can academic departments shift to meet professional needs in these ways?

Notes

1 The authors acknowledge support from NSF grants 2313849, 2137791, and 1923956. The authors also acknowledge the support of the CAHSI organization and the CAHSI backbone.

2 Hispanic-serving institutions are designated by the US government as institutions that enroll at least 25 percent Hispanic student populations. Institutions that enroll Hispanics at this level are eligible for federal funding to support diversifying the college-educated population. Funding is based on numeric representation, rather than acts of serving student populations.

3 The name used for the student described in the vignette is a pseudonym.

4 The authors use the terms Latine, Hispanic, and Latino/Latina throughout this chapter. The Latine term is used to acknowledge nonbinary elements of gender in referring to individuals and groups of individuals of a given ethnicity. Hispanic is often used when addressing the population served by federally designated Hispanic-serving institutions (HSIs). In cases where data collection instruments or systems have specified language related to ethnicity (e.g., institutional research offices use "Hispanic") we use the same terminology.

5 The Great Minds in STEM conference is presented by the Hispanic Engineers National Achievement Awards Conference (HENAAC) organization. The conference was originally titled HENAAC but was rebranded in the last decade. In this chapter we refer to GMiS for clarity.

6 GMiS and CAHI received the Excellencia in Education award (https://www.edexcelencia.org/what-works/examples-excelencia).

7 Present tense is used in this section to reflect the continuous nature of the relationship.

8 CAHSI institutions have traditions of serving first-generation students who also identify as Hispanic or Latine.

9 The external evaluation team regularly obtains institutional review board approval for data collection in the CAHSI effort.

10 Eighty-four student participants completed the survey out of 275 who received the instrument (31 percent). Of those who responded to demographic data, 88 percent were Hispanic/Latine, 29 percent self-identified as women, 67 percent self-identified as men, and 4 percent preferred not to state their gender on the survey. Respondents came from regional state schools that are designated HSIs as well as research 1 and research 2 HSIs.

11 Four choices were provided for the career development item ("not at all supported," "somewhat supported," "supported," and "greatly supported").

12 These items were Likert scales-reported percentages collapsed across "agree" responses (somewhat agree, agree, and strongly agree).

13 Selfie check-ins involve students sending an image of themselves to their college or university staff chaperones at a given time.

References

Espinosa, L., K. McGuire, and L. M. Jackson, eds. 2019. *Minority Serving Institutions.* National Academies Press. https://doi.org/10.17226/25257.

Garcia, G. A., A. M. Núñez, and V. A. Sansone. 2019. "Toward a Multidimensional Conceptual Framework for Understanding 'Servingness' in Hispanic-Serving Institutions: A Synthesis of the Research." *Review of Educational Research* 89 (5): 745–84.

Gates, A. Q., P. J. Teller, A. Bernat, N. Delgado, and C. K. Della-Piana. 1999. "Expanding Participation in Undergraduate Research Using the Affinity Group Model." *Journal of Engineering Education* 88 (4): 409–14.

Gutiérrez, K. D. 2008. "Developing a Sociocritical Literacy in the Third Space." *Reading Research Quarterly* 43 (2): 148–64.

Hamblin, L., D. Barker, and V. Arghode. 2020. "A Phenomenological Approach to Explore Faculty Perceptions about Invisible Labor." *Community College Journal of Research and Practice* 44 (10–12): 804–18.

Hanleybrown, F., J. Kania, and M. Kramer. 2012. *Channeling Change: Making Collective Impact Work.* FSG.

Hug, S. 2018. "Developing Communities of Practice to Serve Hispanic Students: Supporting Identity, Community, and Professional Networks." Paper presented at 2018 ASEE Annual Conference and Exposition, Salt Lake City, Utah, June 2018. https://doi.org/10.18260/1-2–30305.

Hug, S. 2023. "Shifting Identity Trajectories within a Scholarship Program: Local Community Practices that Shape Computing Careers." Paper presented at 2023 ASEE Annual Conference and Exposition, Baltimore, Maryland, June 2023. https://doi.org/10.18260/1-2—44216.

Hug, S., S. Jurow, and H. Thiry. 2021. "HSI Students' Experiences of 'Servingness' during COVID-19." *Dialogues in Social Justice: An Adult Education Journal* 6 (2): A1147.

Hug, S., R. Cota, R. C. T. Castillo, E. Pontelli, and A. M. Delval. 2024. "Borderlands First-Generation-in-Engineering Experiences-Learning with and about Students at the Nexus of Nation, Discipline, and Higher Education." Paper presented at 2024 ASEE Annual Conference and Exposition, Portland, Oregon, June 2024. https://doi.org/10.18260/1-2–48402.

Kezar, A., and E. Holcombe. 2017. *Creating a Unified Community of Support: Increasing Success for Underrepresented Students in STEM.* Pullias Center for Higher Education, University of Southern California, Rossier. https://pullias.usc.edu/wp-content/uploads/2023/10/CSU-STEM_Collab-Report_9617_FINAL.pdf.

Noddings, N. 1984. *Caring, a Feminine Approach to Ethics and Moral Education.* University of California Press.

Núñez, A. M., and R. E. Gildersleeve. 2016. "Sociocritical Matters: Migrant Students' College Access." *Educational Policy* 30 (3): 501–35. https://doi.org/10.1177/0895904814553157.

Núñez, A. M., S. Hurtado, and E. C. Galdeano. 2015. "Why Study Hispanic-Serving Institutions?" In *Hispanic-Serving Institutions*, edited by A. M. Núñez, S. Hurtado, and E. C. Galdeano, 1–22. Routledge.

Rivera, J., and A. M. Núñez. 2022. "Staff at Hispanic-Serving Institutions: Debugging Challenges in Navigating Computer Science." *About Campus* 27 (1): 38–47.

3

Allies Programs in Scientific Societies: Enhancing Meeting Safety and Inclusiveness by Implementing Codes of Conduct and Bystander Intervention Strategies

Paula Brantner and Sherry Marts

Prevalence of Harassment

Despite increases in the number of women in science, technology, engineering, and mathematics (STEM), they continue to face obstacles to advancement. Sexual harassment remains a significant cause of the attrition of women from STEM research, forcing women to either endure toxic work environments or seek employment outside of STEM. Data from surveys of workplace sexual harassment show that academic workplaces are surpassed only by the military in the rate of sexual harassment experienced by employees (Ilies et al. 2006).

As reported by the NASEM Committee on the Impacts of Sexual Harassment in Academia,

> In a survey conducted by the University of Texas System, about 20 percent of female science students (undergraduate and graduate) experienced sexual harassment from faculty or staff, while more than a quarter of female engineering students and greater than 40 percent of medical students experienced sexual harassment from faculty or staff. The Pennsylvania State University System conducted a similar survey and found similar results, with 33 percent of undergraduates, 43 percent of graduate students, and 50 percent of medical students experiencing sexual harassment from faculty or staff.
>
> (NASEM 2018)

While there is a legal definition of harassment that is actionable in the courts for both workplace harassment (Title VII) and educational harassment (Title IX, Title VI), the definition can be inadequate in many ways. The legal definition requires analyzing whether the conduct is "severe" or "pervasive." When looking at the court cases that have interpreted these terms, it quickly becomes apparent that these standards have been developed to mitigate organizational liability rather than playing a significant role in prevention.

The severe requirement has historically been satisfied by only the most egregious incidents, such as sexual or physical assault or unwanted groping. The pervasive requirement is defined by multiple incidents taking place over a continuous period of time, rather than what are considered isolated incidents. There was one workplace harassment case[1] that was notorious for its "one free grope" rule, where a woman whose coworker reached into her blouse to grope her bare breast was denied relief on the basis that this egregious one-time incident was not sufficiently severe or pervasive. While this case represented one of the most extreme interpretations of the "severe or pervasive" conduct standard—and led to a change in California law clarifying that a single incident could be sufficient to create a hostile work environment—reported harassment cases are replete with instances where conduct we now recognize as tell-tale signs of the intent to harass is not considered severe or pervasive enough to warrant legal consequences.

Types of Harassment Prevalent in Professional Societies

In reviewing existing codes of conduct among professional societies, they typically include provisions intended to address the most prevalent types of harassment that have had the most negative impact within the organizations. Four of these provisions include (i) gender-based violence and harassment, (ii) harassment based on other characteristics, (iii) bullying, and (iv) political expression and free speech.

Gender-based Violence and Harassment

Perhaps the first type of harassment that people think of when they hear the word "harassment" is sexual harassment. The use of the word "sexual" primes us to think that only harassment grounded in sexual interest or that involves sexual touching or assault violates the code. Thus, the term "gender-based violence and harassment" (GBVH) has become more prevalent to describe the range of harassing behaviors that arise from gender bias.

One of the key documents to use this term is Convention C190, the Harassment and Violence Convention, adopted in 2019 by the International Labour Organization (ILO).[2] The ILO convention defined GBVH to mean violence and harassment that is directed at persons because of their sex or gender or disproportionally affects persons of a particular sex or gender and includes sexual harassment. GBVH also encompasses any form of disparate treatment based on sexual orientation and gender identity. Whether an organization spells out those types of identity or considers them to be covered more broadly under the GBVH umbrella, it is one of the primary forms of harassment encountered within society spaces.

An internet-based survey of meeting attendees conducted in 2015 found that harassment at meetings included comments on appearance and other forms of verbal harassment, unwanted touch, stalking, and sexual assault. Targets[3] of harassment reported that the experience of being harassed caused them to think more about their personal safety at meetings, avoid social events, and stop attending meetings where harassment occurred (Marts 2016).

Harassment Based upon Other Individual Characteristics

Even though most codes of conduct focused on GBVH, other groups were also subjected to harassing conduct. Race and religion are the most common individual characteristics targeted, but age and disability are also commonly singled out characteristics. When creating a code of conduct, organizations have the opportunity to designate a wide range of individual characteristics as protected from harassment and discrimination. Organizations can also customize their codes to reflect particular forms of harassment and discrimination that are unique to their organization, such as those prohibited in the jurisdiction in which they exist or groups of members with different identities who participate within their organization that may experience different treatment, such as educational level or profession.

Bullying

One way that organizations can distinguish themselves in their codes of conduct is by including anti-bullying provisions. Bullying that is not based on the individual characteristics identified above is typically not prohibited by US workplace discrimination laws. Yet bullying is known to be a significant problem in the academic environment and can readily happen in in-person interactions at conferences. In meetings that feature research presentations, it is not uncommon for audience members to ask hostile questions and belittle the presenter, sometimes until the presenter suffers a significant loss of composure or begins to cry. While a robust exchange of scientific views is expected and encouraged, organizations want presentations to remain civil and not be monopolized by a hostile audience member who questions the validity of the presenter's competence and intelligence.

Organizations are also concerned with bullying behavior in their members' home institutions. Early career professionals are very vulnerable in siloed academic environments where an individual's entire field of research and career advancement depends on the support of those within a highly niche area of focus. When a handful of senior academics can determine the ability of their students to pursue research or teaching in the student's area of interest, their fate can be determined by whether their supervising professor is a mentor or a bully. Those who experience bullying far and above the typical forms of interaction in the academic environment may be fearful of retaliation for speaking out about their experiences, may experience delays in their career after transitioning to a different supervisor, or may choose to leave the field entirely if they are inadequately supported by the small number of individuals in their area of study.

Political Expression and Free Speech

In an increasingly polarized political environment, differences of opinion may play out within organizations and particularly in meeting settings where members interact more directly with each other in person. In 2024, society leadership and meeting planners were particularly concerned about political discord heating up before the US elections in November 2024. Scientific societies were already grappling with differences among members on such hot-button topics as abortion; Covid-19 vaccines; diversity, equity, and inclusion initiatives; and even evolution. While fears about increased conflict at 2024 annual meetings were mostly unfounded, some recent isolated incidents could be attributed to differing political views, which included:

- Objecting to a researcher's inclusion of the words "Free Palestine" on a presentation slide.
- An individual with creationist views who singled out a researcher for their focus on evolutionary science.
- Questioning the organization's judgment in holding an in-person meeting during the outbreak of a new Covid-19 variant and attacking the organization's leadership on social media.
- Hostile questioning of a presenter who discussed DEI initiatives at their home institution.

Given that much of the American election-related polarization remains following the 2024 elections and that specific issues such as the Israel–Palestine conflict have worldwide implications, how members express themselves politically within scientific societies is likely to remain an area of concern that organizations will need to effectively address within their policies and conflict response.

Role of Scientific and Professional Societies in Addressing Harassment

Participation as a member of a scientific society is an important component of a scientist's continued professional learning and career advancement. As acknowledged by the National Academies of Sciences, Engineering, and Medicine (NASEM),

> Professional societies exist to advance and support their specific disciplines and communities. They often have mission statements and principles that encompass goals for their science, engineering, or medicine and the ethics of their profession, created by their members. Through journals, media, conferences, workshops, student programs, and professional training, they are a powerful influence and important part of the career and advancement of those in science, engineering, and medicine.
>
> (NASEM 2018, 160)

As such, professional societies have the potential to be a powerful driver of change through their position to help educate, train, codify, and reinforce cultural expectations for their respective scientific, engineering, and medical communities (NASEM 2018, 160).

The harassment so prevalent in science, engineering, and medical disciplines in general spills over into the activities of a professional society, particularly at meetings where practitioners of a particular discipline regularly gather. Attendance and presentation at professional society meetings[4] are essential for advancement in scientific, medical, academic, and scholarly careers. Full participation in those meetings includes presenting and discussing research results at posters or platform and plenary sessions; attending meals, receptions, networking, and social events; going on field trips and tours; and participating in alumni and special interest group activities. Some professional societies also conduct job fairs or other formal hiring activities at their meetings, and meeting attendees may also go beyond networking to hold informal interviews with potential research collaborators or graduate students for PhD program slots.

Full participation in meetings requires an assurance that all participants are made welcome and kept safe from unwanted attention, bullying, and harassment. An essential part of addressing harassment in academic science, engineering, and medicine in higher education is ensuring that the same standards of behavior are also upheld in off-campus environments (NASEM 2018, 160). Upholding these standards does not naturally happen but requires deliberate effort by societies to communicate those standards to their members and meeting participants and provide a mechanism by which attendees can communicate with the society to report potential violations.

It is important to remember that participation in meetings, like membership in a scientific society, is a privilege not a right. In addition to the registration fees, the cost of that privilege is respectful, professional, and civil behavior toward fellow attendees. Harassers and bullies do not just abuse their targets. They abuse the privilege of being part of a valuable professional event, their professional community's goodwill, and their colleagues' trust.

As NASEM has acknowledged, professional societies have more freedom to develop independent policies and practices for dealing with sexual harassment than federal agencies do, so they are in an ideal position to take action in preventing sexual harassment and effecting cultural change (NASEM 2018, 160). This is particularly true in the current US political environment, where federal agencies have been instructed to change long-held agency positions and/or ignore existing legal precedent. Workplaces and educational institutions are still nonetheless constrained by the legal underpinnings of applicable laws, which, as discussed previously, may be inadequate. However, employers and institutions are loath to go beyond the bare minimum that statutes require and take a liability-focused approach, rather than a prevention-focused approach. Public employers and educational institutions have complex, constitutionally mandated, due process requirements that make holding wrongdoers accountable particularly time-consuming and expensive. These requirements may discourage reporting due to the perception that the public agency is unlikely to be able to take effective action against the accused individual and that this inaction will put the reporter at risk of retaliation. From all indications, this perception may not be inaccurate in the US during the administration of President Donald Trump, as in the early days of his presidency, he has taken several steps to limit the impact of civil rights laws such as Title IX (prohibiting gender discrimination in education), and Title VII (prohibiting multiple forms of workplace discrimination), and constrict the role of administrative agencies who enforce these provisions (the Department of Education's Office of Civil Rights and the Equal Employment Opportunity Commission respectively).

Organizations Begin Taking a Stand against Harassment

Several societies have come forward in the past decade to take a firm stand on the issue of sexual harassment among its membership, which reflects an increasing awareness that meetings, in particular, are a breeding ground for harassment among members.

In the mid-2010s, incidents of harassment at meetings began surfacing in blogs and were addressed in editorials and opinion columns in journals (Wood 2015). In 2011, a group of women from the open technology movement founded the Ada Initiative (named for Ada Lovelace, www.adainitiative.org) to counter the open hostility faced by women at geek conferences and conventions, including technology industry conventions, free and open-source software conferences, gaming conventions, and comic book and science fiction fan conventions.[5] The Ada Initiative quickly developed an open-source conference anti-harassment policy, and they mounted a successful campaign to promote the adoption and enforcement of such policies at major technology and fan conferences and conventions.

At about the same time, reports of harassment at scientific meetings began to emerge in science blogs. In 2013, an anonymous guest blogger for the *Women in Astronomy* blog described an encounter with someone she described as a "known serial harasser" (Johnson 2014). The same blog published an account of harassment at a meeting of the American Astronomical Society (Schmelz 2015). The resignations of Geoff Marcy from his position at the University of California, Berkeley, and Brian Richmond from the American Museum of Natural History increased the visibility of sexual harassment at conferences (Balter 2016; Witze 2016). Among the most egregious incidents in both of these cases occurred at professional conferences. The "Survey of Academic Field Experiences" published in 2014 further documented the extent and type of off-campus harassment that women in science experience (Clancy et al. 2014).

Blog posts with advice on what to do about the problem also appeared. Most of the advice offered assumed that women should not expect conference producers (primarily scientific societies) to address sexual harassment at their meetings and that women needed to develop strategies for handling harassers and harassment on their own (Singer 2012). In a 2015 blog post academic editor Hilda Bastian flatly stated, "There'll be inappropriate jokes and demeaning, unprofessional conversations. Women will be harassed – especially at social events" (2015).

Eventually, professional societies began to follow the lead of the Ada Initiative and adopt codes of conduct for their meetings. To date, hundreds of conferences have now adopted a code of conduct or other anti-harassment policy, and nearly 20 major conference sponsors now require a policy for all sponsored events (Geek Feminism Wiki 2019). In 2022 the NSF updated its meeting funding policy to require that NSF-funded meetings have

> a policy or code of conduct that addresses sexual harassment, other forms of harassment, and sexual assault, and that includes clear and accessible means of reporting violations of the policy or code-of-conduct. The policy or code-of-conduct must address the method for making a complaint as well as how any complaints received during the conference will be resolved. This policy or code-of-conduct must be disseminated to conference participants prior to attendance at the conference as well as made available at the conference itself.
>
> (NSF 2021)

These requirements are all part of best practices for implementing codes of conduct at meetings, but some groups have taken further steps to strengthen their code of conduct enforcement by implementing an allies program.

History of Allies Program Development

As more harassment incidents at meetings emerged, women members of at least two scientific societies took action. It began with Astronomy Allies, a group of women members of the American Astronomical Society who were tenured faculty, secure in their careers. They recognized that by taking on the role of ally to those who experienced harassment, they were putting themselves at risk of retaliation by their colleagues. The risk of retaliation for those in research careers is significant, given the ease with which colleagues who serve as peer reviewers, conference chairs, and promotion and tenure committee members can undermine a career with little or no accountability.

The Astronomy Allies, and other allies groups, comprise society members who volunteer to serve as sympathetic listeners for meeting participants who have experienced harassment, bullying, or other unwanted behavior. They commit to keeping these conversations strictly confidential. They may help targets think through the decision to report code-of-conduct violations. Allies are identified by a highly visible badge, button, or ribbon. (Note that this means volunteer allies may choose to self-identify only when they are fully available to offer help.)

Allies also serve as a deterrent to behaviors that violate the code of conduct. Simply by walking through the meeting, attending sessions, and being present at social events, allies are a visible and physical reminder that harassing and bullying behaviors are likely to have consequences for the harasser.

The Astronomy Allies and other conference allies groups that formed based on this model serve essential roles in the effort to eliminate harassment and bullying at meetings. They provide a safe space for targets to talk about what happened and decide on their next steps. They also reduce vigilante behavior that often arises when meeting participants hear about an incident and assume the meeting staff are not responding, or at least are not responding quickly enough, to curb the behavior.

Incidents of vigilante behavior typically involve a meeting participant who decides to confront the code-of-conduct violator, either in person or on social media. While they may think they are doing the right thing and acting as an ally, often they end up violating the code of conduct themselves. When allies are present, the urge to do something is more likely to take the form of speaking with an ally (and possibly being reassured that a report is being acted on) rather than staging a confrontation with the violator.[6]

The NASEM specifically acknowledged the work of Astronomy Allies in its groundbreaking 2018 report.

> Members of AAS [the American Astronomical Society] have also developed grassroots efforts to prevent and respond to sexual harassment at meetings. For example, the Astronomy Allies is a self-organized group that serves as a visible resource at conferences to discourage harassing behavior, for example, by offering conference attendees a safe escort back to hotel rooms at night, and offers support and counsel to targets of sexual harassment.
>
> (NASEM 2018, 162)

Ento Allies at the Entomological Society of America (ESA) meetings was one of the earliest allies groups organized by women participants. Gwen Pearson, an entomologist and science educator, started Ento Allies after she had become known for speaking out about harassment and several younger scientists approached her at an ESA meeting to talk about incidents of harassment they had experienced. Founded in 2014, the group's development paralleled work by ESA staff and volunteer leadership on their conference code of conduct and its enforcement. Initially, Dr. Pearson reached out to members of the Ada Initiative for advice based on their experiences. She intended to form a small group of experienced scientists who regularly attended ESA meetings. These Ento Allies members, easily identified by their "Ento Ally" badges, would be available to listen to those who wanted to talk in confidence about their experiences and, if asked, offer options for responding. Ento Allies dealt with four incidents of harassment at their first ESA meeting, including one that involved a faculty member who had harassed Dr. Pearson when she was an undergraduate.

As they became better known, Dr. Pearson was sometimes asked to serve as what she calls a "PPE—Personal Protective Entomologist" for participants whose harassers (some of whom were under-restraining orders at their institution) were present at the meeting. In this role, she would accompany the target to presentations and events and simply be beside them to signal that the target was not alone.

After ESA adopted and began enforcing its code of conduct, Ento Allies remained independent of the official reporting and enforcement process. However, ESA staff, particularly Chief Operating Officer Rosina Romano, worked alongside Ento Allies to share information and facilitate incident reporting.

From the beginning, Ento Allies members were experienced academic members who had tenure or were otherwise insulated from retaliation or career damage for doing this work. The original contingent was people Dr. Pearson knew personally and who had themselves been targets of harassment. As the group became better known many ESA members, including students and postdoctoral fellows, asked to join. Dr. Pearson and the original Ento Allies decided to keep the group small and limited to members with well-established careers—one member is a former ESA president. Their vetting process is informal and focused on ensuring that Ento Allies understand the impact of harassment on the target, are willing to listen without immediately offering advice or solutions, and understand that Ento Allies exists to support targets, not to enforce the code of conduct or punish harassers.

At the start, most Ento Allies members had received active bystander intervention training at their home institutions. Within a few years, ESA began providing that training, as well as training in trauma-informed response, as part of their anti-harassment efforts. All ESA staff at meetings received active bystander intervention training and those involved in code-of-conduct enforcement received trauma-informed interview training as well. While the earliest allies groups were self-organized and had minimal or no formal training for the role, most societies with allies groups now offer training in trauma-informed responses and active bystander intervention.

Other societies took notice and formed allies groups, including the American Geophysical Union, the Evolution Meeting, and the Geological Society of America. These groups were organized by the society rather than by members acting independently but were independent of the official code-of-conduct reporting and enforcement processes. Other groups, including the Animal Behavior Society, the Society for Molecular Biology and Evolution, and the Society of Integrative and Comparative Biology (SICB), have also developed allies programs with the assistance of the coauthors. SICB is one of multiple scientific societies managed by Burk & Associates (www.burkinc.com/), a scientific association management and event planning firm, who ensures their clients with annual meetings have instituted allies programs as well as trained leaders and members.

Meeting planners for other types of organizations are taking cues from what has been developed by scientific societies. *Convene,* a magazine published by PCMA (formerly known as the Professional Convention Management Association), a global community for the business events industry, recently published an article on allyship programs to share some of the successful strategies implemented in scientific societies by the coauthors (Dienst 2023).

Allies Programs as Part of a Comprehensive Code-of-conduct Enforcement Mechanism

Allies programs are most effective when developed as part of a comprehensive code-of-conduct implementation system with the following six integrated components: (a) the policy itself; (b) training, as discussed below; (c) a reporting mechanism; (d) an investigation protocol; (e) an accountability process for those who have been found to have violated the policy; and (f) ensuring sustainability by funding code enforcement over time and building trust with the existing system. In turn, allies programs provide support and enhance the effectiveness of all six components of the integrated system.

Policy

As part of the allies training, allies become familiar with the existence of the scientific society's code-of-conduct policy, how to access more information about the policy and its reporting mechanism, and can even provide helpful feedback to organizational leadership about perceived gaps and inadequacies or the need for further clarification on particular provisions. If allies are contacted about potential code violations by meeting attendees who have been targeted, they can walk the attendee through how to report and what will happen next if the attendee chooses to report.

Training

As further discussed in this chapter, having a body of trained allies provides valuable internal support and amplifies the work of the society's leadership, staff, and consultants in promoting a climate consistent with the society's values. Not only do the trained individuals have their own personal understanding of what conduct is inappropriate and the capacity to advise others about that prohibited conduct, but they also learn how to intervene quickly and capably to interrupt harassment and prevent its escalation. This reinforces the required bystander intervention trainings at the allies' home institutions by allowing allies to become a more knowledgeable and confident ally and return from the meeting with new knowledge and approaches.

Reporting

Allies reinforce to other meeting attendees that there is an easy-to-use reporting mechanism that can be trusted. By understanding how the reporting system works, they can provide any necessary reassurances to a potential reporting party that their report will be taken seriously, which can lead to accountability and a safer meeting environment. The ally can also convey that reporting can be confidential, or even anonymous if so desired, and provide any requested assistance, such as helping the traumatized individual gather and organize their thoughts and escorting them to meet with the designated individual accepting reports. Absent an ally's encouragement, some reports simply would not be made due to the potential reporting party's fear of retaliation or skepticism about how their report will be received.

Investigations

The ally's role as both a potential confidant and intervening bystander makes it more likely that they will be a witness to potential code violations. For reporting systems that allow reports from bystanders,[7] a witnessing ally may submit a report on behalf of the target. They can facilitate investigations by identifying other potential witnesses and can encourage reluctant witnesses to become involved. When allies are trained to be attuned to what is happening so they may intervene as quickly as possible, their alertness and understanding of the code's nuances can make their involvement as a witness very valuable to the investigator, particularly as a trusted individual who has demonstrated their commitment to the code of conduct and its effective enforcement.

Accountability

Often, allies are drawn from the society's leadership. Some societies limit participation in allies training exclusively to identified leaders, while others open participation to all interested meeting attendees willing to be trained. When a trained ally is part of the decision-making process that provides consequences for accused individuals whose conduct is found to have violated the code, their expertise is valuable to other process participants who may not have been trained or have direct experience evaluating harassment, discrimination, and bullying claims. Their commitment to a harassment-free meeting environment correlates with their desire to hold violators accountable for breaching that expectation and harming society members and meeting attendees.

Sustainability

An allies program cannot be a one-time effort that permits the society to check a box without continuing to monitor the meeting climate and build awareness of the code. Also, due to the transient nature of meeting attendance, the society must devote efforts to training new allies every year to compensate for those who are not registered to attend a particular meeting, are no longer in the field, or are burned out. Even when a trained ally remains continuously involved in the allies program, many allies welcome an annual training refresher. Witnessing the success of the allies program with a critical mass of enthusiastic and visible allies that blanket the meeting makes it more likely the society's leadership will make both a commitment to the society's code and the expectations of meeting attendees it exemplifies. Hopefully this would take the form of an annual budgetary commitment to funding the allies program so that it continues to grow and be embedded in the organization's culture.

Key Components of a Successful Allies Program

Recruitment and Vetting

Societies with active committees or interest groups that support members from historically excluded groups (i.e., women, people of color) find these to be a good source of volunteer allies. In addition, as Dr. Pearson discovered, once an allies program is announced or becomes visible, volunteers appear. When recruiting volunteers, it is crucial to identify individuals who have the following characteristics:

- An understanding of societal privilege and how harassment and bullying reinforce systemic racism, sexism, homophobia, transphobia, and other biases.
- A willingness to acknowledge, apologize, and make amends when their own behavior reveals unconscious biases.

- Are trained in trauma-informed communication and have an understanding how targets of harassment or bullying may act or react to being targeted.
- Are trained in active bystander intervention.

Note that organizations with conference allies programs usually provide these trainings.

Once recruited, volunteers must be carefully vetted to ensure the safety of targets and other reporters. Vetting individuals who volunteer for allies programs can be challenging. Of concern is that repeat harassers may volunteer for the program as a way of concealing their own misbehavior. Additionally, for reasons that are well-documented (Cantalupo and Kidder 2019), societies cannot rely on campus and workplace anti-harassment policies and investigative proceedings to identify most harassers. Thus, the vetting of potential allies often relies on anecdotal or word-of-mouth information along with self-reporting.

Several of the authors' clients have used a questionnaire that serves as a screening tool and helps ensure that those who wear a conference ally badge have the appropriate beliefs, attitudes, and background to act as an ally. The questionnaire is not anonymous, but the respondents' identities are known only to the independent consultant advising the program.

Questions Covered by the Questionnaire

- The reasons why the respondent wants to serve as an ally.
- What previous training or relevant experience they have.
- Whether they have ever reported harassment or bullying to a supervisor or a Title IX or human resources office as well as the outcome(s) of that report(s).
- Whether they have ever been told their behavior was discriminatory or unwelcome and how they responded.
- Whether they have ever been formally accused of discriminatory behavior, harassment, or bullying as well as the outcome of the investigation, if one took place.

During the authors' consultation, if a prospective ally indicated that they had been formally accused of discriminatory behavior, harassment, or bullying, the independent consultant followed up to learn more about the source of the report and the outcome. This is necessary because such reports are often filed by individuals accused of harassment against targets as a form of retaliation and are found to have no basis.

This vetting process is less than ideal and requires careful follow-up to ensure that confidence in the allies program remains high. Should an ally be accused of behavior that undermines the confidence in the program, how that ally responds is often indicative of their fit to continue serving as an ally. Good allies will have the necessary understanding of how their behavior may reflect their own privilege or unconscious bias and the willingness to apologize without defensiveness when they inadvertently cause offense.

In one instance handled by one of the coauthors, the ally accused of harassment was utterly unaware of what conduct they engaged in that potentially violated the code. Upon further investigation, it was based upon conduct that had previously been independently investigated and found not to have violated any scientific ethics principles as alleged. When the home institution indicated no evidence of an ethical violation, a harassment complaint against the ally was then lodged with the home institution. That complaint was ultimately found to have been retaliatory in nature because the original complaint was not sustained. The ally was temporarily removed from active ally duties while the organizational report was being investigated but ultimately returned to the allies program after a credible witness from the home institution indicated that all of the complaints were retaliatory. This demonstrates the importance of carefully investigating all reports to ensure that allies receive the same consideration under the reporting system that all other meeting attendees would receive as parties and have not engaged in disqualifying behavior.

Active Bystander Intervention Training

Active bystander intervention trainings as a violence intervention strategy at academic institutions became increasingly common in the early to mid-2000s. One of the most commonly used approaches, Green Dot, was developed by Dorothy Edwards after she was appointed director of the University of Kentucky's Violence Intervention and Prevention Center in 2005 (Cherkis and Grim 2015).

The Green Dot, and similar trainings, specifically focus on undergraduate populations and on the prevention of sexual violence in situations where students commonly find themselves. Similar workplace trainings have been shown to "change social norms and empower students to intervene with peers to prevent assaults from occurring" (Feldblum and Lipnic 2016).

More recently, trainings that address how to recognize more subtle aggressive or exclusionary behavior that may be based on gender, race, or other personal characteristics have become more common. A study published in 2021 found that a single two-hour training for university faculty raised participants' confidence in their ability to address subtle acts of exclusion and raised the likelihood that they would intervene (Haynes-Baratz et al. 2021; Kuhn et al. 2023).

Trainings that address both obvious and more subtle forms of harassment, bullying, and exclusion are critical to an effective conference allies program. Conference ally trainings should include how to recognize problematic behaviors; how to intervene without escalating the situation; and how best to communicate with those who have experienced or witnessed trauma. The latter should emphasize the importance of allowing the target or witness to decide whether to involve security, police, or other authorities and whether to report the incident as a code-of-conduct violation.

Given the power disparities that may exist at conferences, where early career professionals are sharing their research (often for the first time) through presentations and posters, hoping to connect with senior scientists in the field to land spots in PhD programs and research labs, it is also important to include information about the role of power dynamics in harassment during training (Kuhn et al. 2023).

If the conference schedule allows, offering training on active bystander intervention to all conference participants is an effective way to boost the impact of a conference allies program. The overarching messages of good active bystander intervention trainings are awareness, "This is what unwelcome behavior looks like;" permission, "Yes, you may intervene when you witness something;" and skills, "Here are effective and proven techniques for intervening."

Building Awareness through Integration with Meeting-related Communications and Marketing

Once ally program participants have completed their training, the work begins before the organization's meeting to ensure that meeting participants are aware of the allies' role before any incidents occur at the meeting and know that they may seek support and assistance from allies if needed.

Besides having allies wear pins or ribbons that signify their participation during the meeting, Burk & Associates' Jill Drupa tries to generate awareness of the allyship program through the event website and any messaging sent out in advance (Dienst 2023).

Among the many ways that societies create awareness of the code of conduct, its reporting mechanism, and the allies program are:

- *Recruitment efforts before the meeting*—Solicitations to participate in the allies program, including explanations of the allies' role and the ability to enlist support during the meeting, raise awareness, even among those attendees who choose not to participate in the allies program.
- *Targeted recruitment efforts*—Efforts to boost participation among society leadership, program committee members, moderators, and speakers demonstrate the society's firm commitment to the allies program when prominent members participate.
- *Meeting website and registration materials*—Details about the allies program are included in the information shared and distributed to meeting registrants.
- *Links and quick-response (QR) codes*—Customized links and QR codes are made widely available to attendees and shared in advance with allies so that all meeting participants can quickly access the code of conduct and reporting mechanism.
- *Slides and announcements*—Allies information and the significance of the buttons or badges are shared during the welcome and announcements at the beginning of the meeting, and the background slides are reviewable by participants while waiting for the meeting to start. It is particularly effective to have the society's president or chief executive officer share information and communicate the importance of the code and the presence of allies.
- *Meeting apps or microsites*—For organizations that use a mobile app or microsite to contain all information that registrants might need during the meeting, information about the code of conduct and allies can be integrated; the most effective placement is near the Wi-Fi password or other critical information.
- *Postcards or fliers*—Written communication that references the allies program and identifying information is distributed at the registration desk and exhibit hall.
- *Signage*—Signage near the registration desk, exhibit hall, and social events with alcohol (along with links and QR codes for the code and reporting form as well as allies information) makes the information accessible to attendees when it is most likely to be needed.

- *Daily attendee communications*—For those meetings that have daily updates for registrants, the allies' presence is reinforced daily by including information about the code and its reporting mechanism.
- *Reporting form*—In the reporting form made available to attendees, potential reporters are alerted to the presence of allies to provide support and a discussion of options, particularly when the individual(s) to whom the report is sent are not on-site or may not be immediately available.

Drupa concludes, "I think [having an allyship program] has really helped change the dynamic of how meetings feel," adding that it helps to communicate "that we're just taking care of everybody" (Dienst 2023).

Case Study of an Allies' Program Effectiveness

One coauthor has worked for three years with a scientific society that has wholly embraced the allies' program and integrated it with its preparation for the organization's annual meeting for the last three years. Before the first implementation of the allies program in 2022, the society had swiftly and publicly responded to an in-depth news article that identified a prominent member of the society engaged in sexual misconduct at his home institution after engaging in similar misconduct at a prior institution. This member had overseen the society's fellowship program for a decade and thus wielded power over early career professionals. As a result of initiating an investigation against the member, the society reached out to multiple survivors of the member's misconduct and provided support by affirming their experiences and giving reassurances that the member would not participate in future meetings and would face consequences for his actions.

The public nature of this incident also reinforced the need for a robust code of conduct and reporting process, which the society had coincidentally adopted just before the news article broke. This incident was the first test of that process and resulted in a lifetime ban for the member after they declined to participate in the investigation. During the investigation, one witness interviewed revealed having shared information about this member's inappropriate behavior back in 2017. However, due to the lack of a code enforcement process, the report was filed away and never acted upon.

When the allies program was first introduced before the 2022 annual meeting, interest was high, and approximately 40 allies (over 5% of the meeting's attendees) participated in the first-time program. Although the training was originally intended to be offered in person, due to the presenter's last-minute unavailability, it was offered via Zoom immediately before the meeting. Allies program participants who could not be present for the live presentation were given access to a training video and affirmed that they had reviewed the video before being given their allies badge. Only a few did not follow through with their initial commitment to the program, most of which were due to scheduling conflicts, travel delays, or their inability to attend the conference.

That first year, multiple reports were submitted through the reporting system, and one ally specifically provided support on-site to an attendee who experienced conduct that likely violated the society's code of conduct. Even though this attendee ultimately chose not to use the reporting system to move forward with an investigation, having a trained ally present that was familiar with the society's code and knew to consult with the safety officer (one of the coauthors) meant that the attendee knew the options available to them and that they were supported by the organization should they choose to move forward. The other submitted reports led to changes in two presentations, where offensive images shared during in-person presentations were removed before the presentations became part of the official meeting record and were shared with virtual annual meeting attendees. While it was unclear whether the additional reports involved allies, the submission of reports demonstrated awareness of the first-time reporting process used to implement the society's updated code of conduct.

In the second year the program was offered, there were approximately 55 allies, which was remarkable because about half were new participants in the allies program. Those who had received the training the previous year were permitted to return as allies and offered a refresher training if desired. However, due to the location of the first year's conference compared to the second (on a different continent), a significant number of previously trained allies were unable to attend the annual meeting the second year of the allies program. This year, the allies program training was offered virtually approximately a month before the program. This was a more optimal time for attendees since they did not have to juggle competing meetings on-site at the annual meeting nor worry about arriving at the conference before the first full day. Nor did they have to watch the training video once they were on-site, which previously added additional stress to allies who already had a full on-site schedule.

This second year's annual meeting was remarkable for an on-site incident in which this society's allies were extensively involved. The organization's annual meeting was moved at the last minute from another state in the United States due to recently enacted laws that specifically targeted LGBTQ+ individuals. The society determined that meeting in that state would be contrary to the organization's values of inclusion and made arrangements to move the meeting to a city and state considered more hospitable to all of the organization's members.

Unfortunately, that was not the actual experience at this annual meeting in the new and presumably more welcoming location. The meeting space was shared with a religious organization that was explicitly hostile to the rights of LGBTQ+ individuals and had meeting signage with negative references to transgender individuals. The groups were assigned to the same section of the large convention center space, which meant the two groups took their breaks in the same corridor and shared restroom spaces. Multiple society members were the target of hostile remarks based on their sexual orientation, gender identity, and appearance. While the society's leadership quickly convened to alert the meeting space's staff and to issue a statement decrying the hostile treatment, there was little that the meeting space's staff could do to limit interactions between the individual participants of the competing meetings given that both groups were contracted to use the space.

Allies jumped in to convene an LGBTQ+ caucus meeting where attendees who had already been targeted or feared being targeted for their identity could share their experience, provide mutual support, and accompany each other to spaces where they were likely to interact with participants from the other group, such as in restrooms. Allies met with individuals who had been targeted and assisted them in filing code-of-conduct reports. Although the code-of-conduct's reporting system obviously had no jurisdiction over individuals who were not members of the society, the meeting space's staff had solicited reports of all relevant incidents. The hope was to create a record that might make the religious organization unwelcome to contract the space in the future.

Having approximately 7 percent of the meeting's attendees wearing allies ribbons, some of whom were also wearing positive LGBTQ+ buttons, created an atmosphere where those who were targeted and felt vulnerable knew they were surrounded and supported by the organization even in an unexpectedly hostile environment and that there were sympathetic individuals who they could approach to discuss a particular incident. Following this annual meeting, the society changed the language in its annual meeting contracts and was now aware of the need to identify groups that might be sharing the space in advance. The society also collected enough reports to effectively prevent the other group from being permitted to return to that venue. Having allies playing a pivotal role in providing support and encouraging reporting that led to a positive outcome in an unexpectedly harmful situation: a situation that the society needed to effectively manage to support its members and adequately reflect its inclusive values in the organization's response.

In year three of the allies program, there were approximately 70 new and returning allies that followed the same vetting and training procedures as previously. This number ultimately represented around 10 percent of meeting attendees. Going into the meeting, the organization had already received and was conducting investigations into multiple reports of incidents at members' home institutions. The society's code of conduct is not limited to incidents that occur in conjunction with the annual meeting but also applies to any conduct between society members that would violate the code's enumerated prohibited behaviors. One of the reporting parties had been part of the initial society team that implemented the allies program in year one, so they were familiar with the process used to report and investigate potential violations. The allies program also plays a role outside meetings to encourage reporting when a member's conduct at their home institution warrants further examination.

At this third meeting, no new incidents were reported during the meeting itself, which was attributed in large part to the critical mass provided by having so many trained allies. Having so many meeting attendees serve as the eyes and ears of code implementation during the meeting is likely to have played a role in encouraging attendees to be mindful of their behavior and suppressing conduct likely to violate the code. Also, having this number of trained allies, some of whom had now served for three years, reflects the society's enduring commitment to the program, where meeting attendees personally witnessed the program's growth and experienced the availability of support and assistance when it might suddenly be needed.

This society is currently in the planning stages for year four of its allies program. It has continued its commitment to a robust code of conduct and has identified necessary changes in the implementation procedures to streamline the reporting process and provide enhanced protection for the parties in the adjudication stage. By publicly addressing the incident in year two, the society has successfully communicated to members its strong commitment to advocating for historically excluded communities to ensure a safe and inclusive space for all at the annual meeting. While identifying new and returning allies for next year's program has not yet occurred, there is every expectation that the program will continue to grow and exceed the 10 percent of meeting attendees designated as trained allies in year three.

Conclusion

Based upon the history and development of allies programs in scientific societies over the last decade as well as the pioneering work of society leaders and allies program trainers and coordinators, having an allies program is now considered to be a best practice and key prevention strategy for harassment, bullying, and misconduct at society annual meetings and within societies more generally.

Successful Allies Programs Reflect the Following Organizational Actions

- Implementing a robust code of conduct of which members are aware and can use to guide their behavior at meetings and within the society.
- Having dedicated and vetted individuals willing and able to provide support and information to attendees that experience unwanted behavior violating the code.
- Training individuals in bystander intervention to effectively respond to incidents in real time and prevent their escalation when appropriate.
- Making allies visible to attendees in advance of and during the meeting itself, with multiple forms of communication and identifiable ribbons or buttons.
- Having a reporting system and investigation process that allies can assist individuals with initiating or use themselves after being trained to spot harassment and discrimination.
- Identifying a critical mass of individuals who can serve as eyes and ears on-site at meetings to discourage inappropriate behavior and are also familiar with the organization's code and reporting process to initiate or assist with reports outside the meeting environment when appropriate.
- Reflecting an enduring and sustainable commitment to eradicating specific forms of behavior in organizational settings as well as taking visible and appropriate steps to do so annually.

For organizations that prize safety and inclusiveness at their meetings and in their organizational spaces, an allies program is an invaluable way to bring those values to life. Such a program will also continually strengthen their organization through a growing and visible critical mass of trained members committed to upholding the highest professional values to which scientific societies aspire.

Notes

1 *Brooks v. City of San Mateo*, 229 F.3d 917 (2000). Incidentally, but probably not coincidentally, the Brooks opinion was written by former Ninth Circuit Court of Appeals Judge Alex Kosinski, who later resigned from the bench following multiple sexual harassment allegations reported by his law clerks.

2 C190—Violence and Harassment Convention, 2019 (No. 190). Adopted in Geneva at the 108th ILC session, June 21. https://normlex.ilo.org/dyn/normlex/en/f?p=NORMLEXPUB:12100:0::NO::P12100_ILO_CODE:C190.

3 "Target" is the word that has become more common in the field generally as a substitute for the word "victim," which reduces agency in our opinion, or "survivor," which is a term embraced by some as part of their identity following assault or harassment, but is not embraced by or identified with by all who have experienced harassment or assault. "Target" is also designed to highlight that this is a chosen course of conduct by the perpetrator or accused.

4 We will use "meetings" in this paper to include conferences, symposia, consortia, and other professional gatherings.

5 A timeline of documented events can be found at http://geekfeminism.wikia.com/wiki/Timeline_of_incidents.

6 An egregious example of vigilante behavior occurred at the 2019 meeting of the Society for American Archaeology (SAA). David Yesner, a serial harasser who had been banned from his own campus, was present at the meeting along with several of his targets. When the SAA took no steps to remove him from the meeting, Michael Balter—a journalist who was at the meeting to present his reporting on sexual harassers in STEM—attempted to escort Yesner from the meeting. SAA subsequently removed Balter from the meeting (Wade 2019).

7 It is highly recommended that societies accept reports from any individuals with knowledge of potential violations. While allies who counsel individual targets should respect their wishes and not personally involve them in a reporting process without their consent, an ally who witnesses a particular incident may have an independent interest in addressing inappropriate behavior and reducing its impact

on others. They may also be aware of a pattern of conduct by the accused of which the most recent target is unaware.

References

Balter, M. 2016. "The Sexual Misconduct Case That Has Rocked Anthropology." *Science*, February 5. http://www.sciencemag.org/news/2016/02/sexual-misconduct-case-has-rocked-anthropology.

Bastian, H. 2015. "7 Tips for Women at Science Conferences." *PLOS Blogs: Absolutely Maybe*, January 18. http://blogs.plos.org/absolutely-maybe/2015/01/18/7-tips-for-women-at-science-conferences/.

Cantalupo, N. C., and W. Kidder. 2019. "Systematic Prevention of a Serial Problem: Sexual Harassment and Bridging Core Concepts of Bakke in the #MeToo Era." *UC Davis Law Review* 52: 2349–405.

Cherkis, J., and R. Grim. 2015. "Here's How You Can Help Stop a Sexual Assault before It Happens." *The Huffington Post*, July 9. http://www.huffingtonpost.com/2015/07/08/bystander-intervention_n_7758118.html.

Clancy, K. B., R. G. Nelson, J. N. Rutherford, and K. Hinde. 2014. "Survey of Academic Field Experiences (SAFE): Trainees Report Harassment and Assault." *PLOS ONE* 9 (7): e102172. https://doi.org/10.1371/journal.pone.0102172.

Dienst, J. N. 2023. "Making Allyship Part of the Program." *PCMA Convene*, May 26. https://www.pcma.org/making-allyship-part-program-harassment-prevention-2/.

Feldblum, C. R., and V. A. Lipnic. 2016. *Select Task Force on the Study of Harassment in the Workplace, June Report.* United States Equal Employment Opportunity Commission. https://www.eeoc.gov/select-task-force-study-harassment-workplace.

Geek Feminism Wiki. "Conference Anti-harassment/Adoption." Last modified May 28, 2019, 21:02. http://geekfeminism.wikia.com/wiki/Conference_anti-harassment/Adoption.

Haynes-Baratz, M. C., T. Metinyurt, Y. L. Li, J. Gonzales, and M. A. Bond. 2021. "Bystander Training for Faculty: A Promising Approach to Tackling Microaggressions in the Academy." *New Ideas in Psychology* 63: 100882. https://doi.org/10.1016/j.newideapsych.2021.100882.

Ilies, R., N. Hauserman, S. Schwochau, and J. Stibal. 2006. "Reported Incidence Rates of Work-Related Sexual Harassment in the United States: Using Meta-analysis to Explain Reported Rate Disparities." *Personnel Psychology* 56 (3): 607–31. https://doi.org/10.1111/j.1744-6570.2003.tb00752.x.

Johnson, J. 2014. "Closing Time at the 'Astronomy Nightclub.'" *Women in Astronomy*, January 6. http://womeninastronomy.blogspot.com/2014/01/closing-time-at-astronomy-nightclub.html.

Kuhn, J-U., N. Arain, J. Bell, B. Davis, H. Kaiser, D. Madden, G. Prepetit, and K. Williamsen. 2023. *Strategies for Developing, Implementing, and Sustaining Sexual Harassment Bystander Intervention Programs for Faculty, Staff, and Graduate Students*. The National Academies Press. https://doi.org/10.17226/27266.

Marts, S. 2016. "Open Secrets and Missing Stairs: Sexual and Gender- Based Harassment at Scientific Meetings." *OSF*, January 27. osf.io/7j9qv.

NASEM (National Academies of Sciences, Engineering, and Medicine). 2018. *Sexual Harassment of Women: Climate, Culture, and Consequences in Academic Sciences, Engineering, and Medicine*. The National Academies Press. https://doi.org/10.17226/24994.

NSF (National Science Foundation). 2021. "Chapter II – Proposal Preparation Instructions." In *Proposal and Award Policies and Procedures Guide*. NSF 22-1. https://www.nsf.gov/pubs/policydocs/pappg22_1/pappg_2.jsp.

Schmelz, J. 2015. "Unreported Sexual Harassment at AAS Meetings: An Example." *Women in Astronomy*, July 6. http://womeninastronomy.blogspot.com/2015/07/unreported-sexual-harassment-at-aas.html.

Singer, K. 2012. "Tips on Anti-harassment Policies – Presentation from DPS 2012." *Women in Planetary Science*, November 5. https://womeninplanetaryscience.wordpress.com/2012/11/05/anti-harassment-policie-dps-2012/.

Wade, L. 2019. "#MeToo Controversy Erupts at Archaeology Meeting." *Science*, April 15. https://www.science.org/content/article/metoo-controversy-erupts-archaeology-meeting.

Witze, A. 2016. "Berkeley Releases Report on Astronomer Sexual-Harassment Case." *Nature*, December 19. http://www.nature.com/news/berkeley-releases-report-on-astronomer-sexual-harassment-case-1.19068.

Wood, B. 2015. "Zero Tolerance. Period." *Science* 350 (6260): 487. https://doi.org/10.1126/science.aad6652.

4

Reflections from the Field: How STEM Society Leaders Navigate Systemic Organizational Dynamics in the Advancement of DEI Goals

Lisa Black, Lucas B. Hill, and Gretalyn Leibnitz

STEM society leaders often face many structural, cultural, and relational barriers in trying to do this important work, which requires careful attention to our own and others' humanity as well as a critical perspective that rejects the status quo and persistent social inequities. In focusing a human-centered approach, leaders must engage in a variety of intra- and interorganizational connections with those who have varying degrees of support and priorities for DEI.[1] These relationships, though often difficult, are vital in making direct and indirect advancements to DEI practices within an organization. Such advancements require a holistic vision of the organization and the heart, grit, and patience to see change occur. Thus, this chapter examines the following key question: *How does a STEM society leader navigate systemic organizational dynamics and relationships to foster inclusive society engagement?*

We work to answer the question by providing an in-depth case study of one STEM society leader's experience advancing DEI within the American Society of Civil Engineers (ASCE). Dr. Lisa Black has served as the director of cultural belonging and social ethos at ASCE since 2020. This case study shares her story in advancing DEI in ASCE where she has attempted to apply a human-centered approach informed by a strategic and systems-oriented lens to DEI change work. This narrative is a heartfelt reflection of her experiences working in an engineering professional society and the relationships that have been fostered to work on persistent under representation and barriers that are manifest in the field of engineering broadly.

This chapter is broken up into three sections. First, we describe persistent under representation in engineering that contextualizes DEI issues for the case study. In this section, we also present two useful lenses for STEM professional and disciplinary society leaders to understand and unpack their DEI reform efforts. Second, we provide a detailed case reflection by Dr. Black that describes her experiences leading DEI-related change in ASCE. Last, we provide recommendations and lessons learned from the case study reflection for STEM society leaders.

Background for the Study

STEM professional and disciplinary societies have a long history of trying to advance DEI within their membership of current and future faculty members, researchers, and practitioners, typically through minority affairs committees (Segarra et al. 2020a). More recently, societies have been tasked with additional collaborative roles in broadening STEM participation such as participating in and leading grant-funded initiatives (e.g., ACCESS and ACCESS+). However, society leaders do not typically receive training in how to lead and navigate intra- and interorganizational connections in pursuit of broadening participation outcomes. Consequently, in this chapter, we specifically highlight the work of Dr. Black, a DEI leader from ASCE. Formed in 1852, ASCE is the oldest national civil engineering society. Currently representing more than 160,000 members of the civil engineering profession in 177 countries, ASCE is also the largest professional society for civil engineers and remains committed to inspiring, connecting, and serving its global membership. The core mission of ASCE is to ensure that the civil engineering profession sustainably advances and protects society's health, safety, and welfare.

Persistent Under representation in Engineering

Studies of engineering suggest a deeply enculturated set of practices and perceptions. Learning engineering culture begins with engineering education (Johri and Olds 2014; Levinson et al. 1996; McIlwee and Robinson 1992; Tonso 1996), which historically has become associated with preferred forms of gender, race, and sexuality (Baillie et al. 2012; Pawley and Tonso 2011; Riley et al. 2014; Tonso 1996, 1999, 2007). Scholars suggest that cultural obduracy has emerged over a long course of history from engineering's tight melding of white, straight, masculinity with technical expertise (Faulkner 2007; Lohan and Faulkner 2004;

Pawley and Tonso 2011; Tonso 2007). Through a tacit set of understandings, engineering students receive recognition and legitimacy by excelling in accepted engineering education practices and social interactions among students as well as between students and faculty (Tonso 2006, 2014). Those given prominence may consciously or unconsciously deploy their standing in ways that elevate historic notions of a white, straight, masculine workforce, while leaving women and other under represented groups feeling suspect as engineers (Baillie et al. 2012; Eisenhart and Finkel et al. 1998; Foor et al. 2007; Hacker 1981, 1989; Tonso 2007).

Women engineers who persist through graduate school and are interested in academic careers, if hired, are immersed in academic environments created over many generations by and for majority men members. Not surprisingly, many research results confirm common STEM academic experiences for women, such as a "chilly" climate (a phrase coined by Hall and Sandler [1982]); criticism for a "lack of professionalism shown" (Trautner et al. 1996); patronization and harassment by students and colleagues; inequitable allocations of work responsibilities; lack of women role models; isolation; hyper visibility, yet invisibility challenges; stereotyping and the resulting stereotype threat; experiences associated with explicit and implicit bias; a lack of recognition and rewards; exclusion from professional or social networks and events; and policies that penalize caregiving responsibilities (Armstrong and Jovanovic 2015; Beede et al. 2011; Bilimoria and Liang 2012a, 2012b; Britton 2017; Cadwalader et al. 2014; Hall and Sandler 1982; Trautner et al. 1996; Trower 2008).

Micro-inequities and micro-affirmations further compound differences experienced by women and men (especially white men) academics. Depending upon one's perspective, micro-inequities are "small covert events, often ephemeral, hard-to-prove, unintentional, frequently unrecognized by the agent, and occur wherever people are perceived to be 'different'" (Rowe 1974, 1975). Micro-affirmations are small acts that "occur wherever people wish to help others succeed" (Rowe 2008). Through incorporating gross as well as micro-inequities and micro-affirmations, engineering academic culture makes and remakes itself in its own image; the lack of inclusivity is produced as normal, and the reasons for engineering culture's makeup seem the fault of those excluded. As such, engineering culture makes it difficult for women to be considered full-fledged members, despite engineering expertise equal to that of men (Faulkner 2007; Riley et al. 2014; Tonso 2007).

Role of STEM Societies

STEM professional and disciplinary societies are in a unique position to accelerate the desired DEI change. Societies may serve as the single professional constant for a STEM-trained individual and consequently, as a critical space for fostering professional identity, belonging, and persistence. This may be particularly important for people from historically excluded groups (Campbell-Montalvo et al. 2022; Smith et al. 2021). Further, based upon the authors' experiences and literature (Leibnitz et al. 2022a, 2022b), societies serve as national melting pots for individuals across various career stages, from a variety of academic institutions, and across different academic and academic-influencing professional domains (e.g., STEM accrediting bodies, national laboratories, corporations, government). Societies also (a) help inform accrediting bodies; (b) set cultural standards for the discipline; (c) model discipline culture of DEI; (d) offer prestigious recognition and awards; and (e) convene national conferences that provide opportunities for broad inclusive engagement and dissemination of information (Leibnitz et al. 2022a, 2022b). All these elements can be leveraged to advance DEI in STEM academic professions. No other entity provides such a capacity to provide career-long support, especially for STEM-trained people from historically excluded groups, *and* impacts STEM culture so broadly.

Role of STEM Society Leaders

STEM society leaders in engineering spaces are entrenched in engineering culture and behavioral attributes. This means that even if society leaders work to support DEI, they may experience tension between the drive to protect the status quo and the societies' history, culture, and accepted inclusion values (Solebello et al. 2016) (see the "Adaptive Challenges Require Adaptive Leadership" section). Legacy ways of being and doing are manifest in implicit society mental models infused in society operations. Society mental models can be reflected, for example, in ideas about what research topics deserve funding or recognition, who is qualified to hold leadership positions, or whose research or scholarship deserves to be highlighted in convenings or society publications. Because society's mental models represent how things are done, these implicit models can create pressure that is counter to desired, even articulated, organizational change. This tension is precisely what Heifetz et al. (2009) defined as an adaptive challenge.

Two Lenses

To further unpack the role of society leaders in advancing DEI, we use two interrelated lenses: adaptive leadership and network theory.

Adaptive Challenges Require Adaptive Leadership

An adaptive challenge can be identified by: (a) a persistent gap between reality and aspiration, (b) an inadequacy of the current repertoire of responses, (c) the requirement of difficult learning, (d) the engagement of new stakeholders across boundaries, (e) its persistence over a longer time frame, and (f) the capacity to navigate disequilibrium as change is enacted (Heifetz et al. 2009). Adaptive leaders need to be willing to diagnose an ineffectual system, challenge the status quo, and courageously pilot change strategies to effect change. Thus, adaptive leadership is "the practice of mobilizing people to tackle tough challenges and thrive" (Heifetz et al. 2009, 14).

Adaptation is reliant upon diversity to advance the evolution of the organization (Heifetz et al. 2009). Consequently, adaptive leaders must engage a broad, diverse group of stakeholders and not do this work alone even if they have formal authority. Formal authority is required; however, it is not sufficient. In professional societies, a DEI committee leader may have some degree of formal authority, having been appointed as a volunteer member to lead the group; or the society might formally employ a staff member to lead society DEI efforts. Ultimately, it is important that the identified leader(s) work with key organizational stakeholders (e.g., see REVIIS in Chapter 5) for successful change.

Adaptive leadership requires that a leader be willing and able to take an organization into and through the productive zone of disequilibrium; namely, pushing people outside their comfort zone, but not so far that they rebel. It requires balancing competing commitments and a willingness to talk about issues that are difficult to discuss. Core identity features of an organization need to be retained, yet others need to be discarded for the organization to thrive in the new changing environment. In professional societies, balancing the tension referenced earlier about implicit models of how things are done (i.e., core identity) may conflict with an articulated interest in broadening participation through the adoption of new recruitment and retention strategies.

For individuals to lead adaptive change they need to balance competing factions; courageously experiment with new strategies; be able to view the organization both from a 30,000-foot level and at the level of the weeds; engage in learning and reflection *before* acting; view and engage in conflict as a means of productive growth; and be willing, and enjoy, making hard choices. A change in the legacy ways of doing society work may risk that seasoned members threaten to, or actually, leave the organization. Adaptive leadership requires active learning and a growth mindset to navigate these tricky spaces. Ultimately, because the process of adaptive organizational change is shared, the adaptive leader needs to be willing to let go of their own definition of success and be able to embrace a potentially unanticipated outcome fostered by the collective. Not everyone chooses to do this type of challenging cognitive and emotional labor; but it can be learned and practiced (e.g., handout). Incorporating theoretical perspectives on systems change, race, and social systems networking is informative for adaptive leaders engaged in DEI work.

Network Theory

The literature is quite clear on the power of social and professional networks in advancing change (Kadushin 2002; Kezar 2011). The success of teams, the ability to maintain competitive advantage within the larger system, and the ability to influence attitudes and behaviors are all contingent on individuals effectively building and leveraging their connections for individual and organizational benefit (Frank and Xu 2020; Joshi et al. 2009; Tushman and Scanlan 1981). Connections can manifest as strong and weak ties (Krackhardt 1992; Granovetter 1983), which indicate various intensities of relationships, both formal and informal, often rooted in formal and informal positionality (e.g., holding a role with organizational authority or being a long-term employee who holds considerable institutional knowledge). These connections range from intraorganizational to interorganizational and from homophilous to heterogeneous, thus creating the possibility of connections with individuals who vary in their organizational memberships as well as demographic, personality, and attitudinal characteristics. Social capital is the most often discussed operationalization of network connections and involves the benefits that accrue for individuals who maintain connections with those with access to resources, power, and behavioral influence on others (Adler and Kwon 2002). The ability to span the boundary between networks, within or between organizations, is a source of considerable social capital since that individual becomes a nexus point for mutual benefit exchange (Burt 1992). Therefore, individuals are (or can become) boundary spanners who can connect and convene constituents in meaningful ways toward change goals (Hill 2020).

STEM society leaders attempting to advance DEI within their organizations already may sit at such nexus points, or have the potential to do so, as they interact with individuals within the core administrative structure of their societies, with members scattered across many different institutions, and with other partners (e.g., grant initiatives, higher education organizations). STEM society leaders may hold different levels of positional authority and informal respect and visibility. Many, if not all, already leverage their connections and relationships in pursuit of DEI reform. However, leaders, not just in the STEM society context, do not often explicitly reflect upon nor map their intra- and interorganizational networks and explore how to strategically increase their efficacy in advancing specific aims and goals. We argue that for STEM society leaders to be successful they must deeply reflect and understand how they develop, maintain, and leverage their connections in strategic ways for DEI reform. In addition, it is vital for this responsibility to not fall squarely on individual leaders but the groups and teams in which they interact most closely around DEI. The distribution and coordination of network leverage, both intra- and interorganizational, increase its power and prevent the accumulation of time burdens that are not sustainable.

Combined Conceptual Framework

Combining the two lenses presented thus far, we present our conceptual framework in Figure 4.1. STEM society leaders are at the heart of the framework and are embedded in both their inter- and intraconnections (i.e., network theory) and bounded by adaptive leadership that operates within the systemic nature of DEI challenges and solutions. In the next section, we will unpack these lenses with a case reflection of the lived experiences of a STEM society leader trying to advocate for and advance DEI reform within their organization.

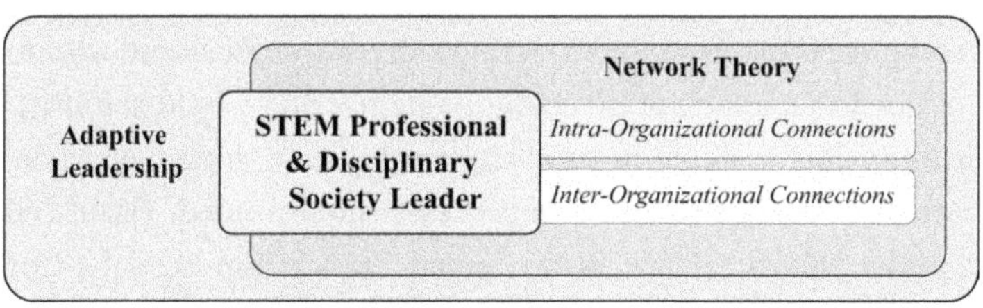

Figure 4.1 A visual diagram of the conceptual framework. Created in Adobe Illustrator by Lucas B. Hill.

Case Study Narrative

For the past thirty years, ASCE has demonstrated a commitment to increasing participation from gendered and marginalized professionals. This commitment has been a major evolution given its history of under representation especially related to women joining the profession. Formed in 1852, ASCE did not allow women into the society until the admission of Elsi Eaves, PE, in 1927. Before that time, Nora Stanton Blatch Barney was among the first women in the United States to earn a civil engineering degree, graduating from Cornell University in 1905. She was also the first woman to earn junior member status in ASCE. In 1916 Stanton Blatch Barney sought to become the first woman to enjoy full ASCE associate member status, but the ASCE Board of Direction (BoD) turned down her application. She challenged the decision in court, but a New York State Supreme Court upheld ASCE. She went on to an illustrious career but did so without having been accorded membership in ASCE, for which she was clearly deserving. Not until August 11, 2015, did the ASCE Executive Committee vote to posthumously advance Nora Stanton Blatch Barney to the status of ASCE Fellow.

Respecting its early history, ASCE recognizes that DEI are mission critical, especially given that the global communities the profession serves comprise people with diverse backgrounds, experiences, aspirations, and abilities. Former National Academy of Engineering President William Wulf (2001, 31), a staunch advocate for increasing diversity in engineering, stated:

> Every time we approach an engineering problem with a pale, male design team, we may not find the best solution. We may not understand the design options or know how to evaluate the constraints; we may not even understand the full dimension of the problem … by failing to attract a diverse engineering workforce, we diminish what engineering can contribute to society, and society pays an opportunity cost.

Wulf recognized the need for attracting a diverse engineering workforce and cogently argued that a lack of diversity in engineering would result in "paying an opportunity cost – a cost in products not built, in designs not considered, in constraints not understood, and in processes not invented" (Wulf 2001, 31).

In the mid-1990s, the ASCE BoD charged a task committee, the Committee on Diversity and Women in Civil Engineering (CDWCE) with researching the topic of recruitment and retention of women in civil engineering. In the late 1990s, the CDWCE evolved from a task force to the Committee on Diversity and Inclusion (CDI) that was charged with focusing on the recruitment and retention of a more broadly diverse community of engineering professionals (e.g., race/ethnicity, age, and ability status). Within the last few years, ASCE leaders have led major policy initiatives such as a major overhaul of the civil engineering code of ethics and the adoption of the justice, equity, diversity, and inclusion-focused policy statement, ASCE JEDI Policy 417 (ASCE 2021).

Fast forward to 2023 when, for the first time in its history, ASCE installed a majority of women BoD members. Leading the board as president, president-elect, and past president are two women and a first-generation person of color who released a call to action to address the disparate number of women and gendered minorities remaining in the profession. As the BoD launched into the strategic planning process, the leaders also integrated DEI into the strategic plan to strategically focus on diversity and inclusion in the civil engineering workforce and higher education environments. These women role models are women who are actively calling into action the expectation that all members will engage in equitable and inclusive practice for the benefit of the profession.

The aforementioned describes the evolution of ASCE's DEI efforts. ASCE aims to lead in building capacity to transform organizational culture. Though the goal of achieving equity and representation of historically gendered and marginalized groups across the organization and profession remains elusive, accepting that such change requires a complex, systemic approach has become realized.

Organizational and systemic change approaches are highly complex technical processes. A caveat is that these well-researched technical processes are driven by human enterprise. Central to professional societies' DEI organizational change efforts are the humans—the STEM professionals and disciplinary leaders intent on shifting the culture. The proceeding is a recounting of the lessons and experiences of the DEI professional navigating as a change agent within ASCE. Outlined are the major observations and experiences that shaped and clarified her path forward and, at times, compelled her to stand still for reflection. What proceeds is neither a detailed rendering of the technical processes nor a step-by-step guide to systems change. Instead, the following reflections are a gift of hope for professional society leaders navigating this space.

Reflection 1: The Power of Passion and Knowledge

My pathway to ASCE as a DEI professional is rooted in my leadership experience in a Midwestern, public K–12 school district where I was responsible for addressing systemic racial inequities in student academic achievement. Those early years taught me two foundational lessons that have transcended the ASCE organizational context: (1) passion, though a powerful motivator, is not enough to address the obduracy of systemic inequities that stymie organizational change; and (2) experiential and technical knowledge are essential to navigating complex landscapes.

Embracing the role of passionate advocate, I was eager to positively impact the social and academic lives of students traditionally marginalized. Shifting from an individual student focus to a systems change agent across forty-eight schools with more than 24,000 students, of which 30 percent were students of color, was a mammoth undertaking that required more than passion. This initial leadership experience forced me to come to terms with the fact that passion with limited knowledge was not enough to prepare or sustain change efforts. My focus became to listen, learn, observe, and lean into the strengths that led me to ASCE.

Intent on understanding the systems that operate and the pivotal role that human enterprise plays within and across systems, I began building my knowledge of ASCE's story through extensive examination of the culture. Included in this cultural exploration was reviewing demographic data, archival documents, and learning about the DEI-related programs and policies. I also met informally with ASCE members who provided insights into the cultural aspects of ASCE. Members identifying as non-white shared experiences and concerns regarding discrimination in the workforce and the lack of professional respect by their white peers. Similar stories were reported by women intent on addressing the retention of women in the profession. I also learned of historic infrastructure inequities that negatively impacted marginalized communities. For instance, the construction of highways through predominantly Black neighborhoods or limited access to clean water in impoverished US rural communities.

A key step in organizing the knowledge gained was to identify a framework to assess and inform strategic actions. With support from the ASCE leadership, we joined the ACCESS+ program, an NSF-funded program whose goal is to prepare and support STEM professional society change leaders. With the goal of assessing the systemic impact of DEI efforts to increase the number of women and gendered minorities in the profession, I administered the Environmental Equity Scanning Tool (EEST; described further in Chapter 12). The EEST revealed areas of opportunity for growth and provided the essential data to inform strategic action planning. The program also facilitated strategic planning efforts for the MOSAIC (Members of Society Advancing an Inclusive Culture) team.

In retrospect, moving beyond passion to building technical knowledge of organizational and systems change proved to be an empowering process. More importantly, it was the experiential knowledge imprinted on the psyche of this DEI professional. The power of those lived experiences is recounted in the next section.

Reflection 2: The Minoritized Professional Leader and Positional Authority

Throughout my career as a DEI professional, I observed that regardless of the organizational context (e.g., higher education, professional society, corporate) and professional preparedness, individuals selected to lead DEI efforts tended to be passionate and non-white. Moreover, the positions were often provided with limited resources in terms of staff support and program funding as well as decision-making authority. The major lessons I share at this point are those stemming from self-awareness and positional authority.

First, self-awareness of intersectional identity within the context of ASCE facilitated my understanding of the systemic barriers that exist for gendered and minoritized civil engineers who rarely see themselves reflected in their workplaces or within their esteemed professional society. Grounded in my lived experiences as a DEI professional, self-awareness empowered me to exercise empathy and impassioned resolve.

Unlike most professional engineering societies, ASCE has invested in hiring full-time, salaried staff professionals to lead its DEI efforts, and while self-awareness serves as a vital internal leadership lens, I am also keenly aware of the historic limitations of these leadership positions. Positional authority is a major challenge. For instance, many who lead in these positions operate programs with no formal staff support and limited funding. Furthermore, the positions do not authorize executive leadership decision-making. I confess that leading from a position with little to no authority to impact systemic change can often feel disheartening. At those moments, I summon my impassioned resolve. Purpose-driven and resolute, I rely on my depth of knowledge, experience, and patience to strategically build a plan of action that advances DEI, while also procuring third-party funds and building collaborative relationships and coalitions to leverage limited resources. Policy statements, educating member leaders, and strategically connecting with and empowering organizational staff have proven impactful in ASCE's DEI journey. The last reflection highlights the value of the partnership network.

Reflection 3: The Power of the Internal and External Village

Facing the inherent leadership challenges that reside in these positions, I learned the single most important lesson—I am not alone in this DEI journey. Focused on fostering a sense of collective responsibility, I began my journey at ASCE building relationships with internal staff and member leaders. For instance, I learned that one of the most powerful internal resources is the vested interest and commitment demonstrated by executive leaders. They have encouraged and supported my initiative and strategic decisions. The power of those relationships continues to sustain my spirit as a passionate, knowledgeable DEI leader.

I also devoted time to learning about the culture of ASCE. Through that exploration, I built relationships with key staff who facilitated the identification of the points of support for strategic action. Recognizing the challenges of positional authority, I created an internal advisory team of leaders and peers who collectively committed to informing and supporting my role in advancing DEI. My internal village also included members of the board-level committee MOSAIC and its predecessor CDI. Working closely with the highly diverse MOSAIC team members affirms my passion and self-efficacy as a DEI professional. They are the co-leaders who inspire my devotion to ensuring their chosen profession is just, equitable, diverse, and inclusive.

Last, external strategic partners such as the National Society of Black Engineers, Society of Hispanic Professional Engineers, Society of Women Engineers, and Women in Engineering Pro-Active Network have been instrumental in sustaining my professional society journey. Connecting and building relationships with my network of professional society colleagues incentivized my purpose and eagerness to contribute my skills and knowledge to achieve the diverse and inclusive cultures we collectively seek. Learning about their programs and initiatives presented opportunities to leverage resources and collaborate on common goals. For instance, ASCE was proud to support and host the launch of the 50K Coalition effort, which brought together several professional societies and academic institutions to significantly increase the number of diverse engineers. The ACCESS+ program is another example of a community of practitioners collaborating and sharing best practices with the goal to increase the number of women and gendered minorities entering and remaining in STEM professions. The camaraderie and collaborative nature of these partnerships have been essential to sustaining my resolve as a DEI professional within the professional society environment.

In sum, the journey of leading the process of organizational and systems change comes with career-defining lessons. Leveraging the power of passion and knowledge to inform a keen self-awareness of the value of intersectional identity is revealed through the lived experience of the professional society DEI leader. The challenge of positional authority as a minoritized professional may be mitigated by building a collaborative network of internal and external partners equally committed to removing the systemic barriers that obstruct fully inclusive participation in the field of engineering.

Despite apparent progress, systemic barriers to full participation in the civil engineering profession for women and members of historically excluded groups persist, and the current sociopolitical climate that attempts to eliminate DEI progress and inclusion is a strong indicator that there is still much work to be done. ASCE has remained steadfast in its commitment to advancing the ideals of DEI, but the marathon journey continues. As ASCE past President Leland J. Walker (1973) wrote over forty years ago:

> The message is beginning to come through loud and clear! If we, as civil engineers, want to retain any semblance of our position as a profession, we must individually and collectively assume leadership to respond actively to the message from the disadvantaged and underprivileged as well as from the advantaged and privileged. The fact that these demands are still made of ASCE and its members is an expression of confidence in the ability and the leadership potential of ASCE.

Recommendations

Based on the lived experiences and lessons learned as a seasoned DEI professional, the following recommendations are offered as encouragement to all professional society leaders devoted to serving as change agents within professional societies.

1. *Lean into an equity mindset*—Central to any organizational and systemic change are impassioned humans who possess their own uniquely diverse backgrounds, identities, and experiences. Embrace your intersectional identity as a leadership strength that provides a window into the experiences of those you serve.
2. *Leverage leadership commitment and support*—Universal to organizations undergoing change is the essentiality of leadership commitment and support. Building trusting relationships with leaders at all levels is essential to implementing meaningful systemic change. Leadership support is also the motivational fuel needed to persevere.
3. *Build a network of support*—Most DEI professionals leading change within STEM professional societies are often navigating under-resourced programs and have a limited capacity to advance strategic, systemic change. Building relationships with internal and external professional peers and leaders can engender a sense of collective responsibility as well as empower partners to leverage resources to advance common DEI goals.
4. *Adopt a growth mindset*—Well-known is the fact that no panacea or silver bullet exists to remedy the historic inimical ills of prejudices such as racism and sexism. Shifting organizational culture is a marathon not a sprint. In tandem with the first two recommendations, prepare for the marathon by adopting a growth mindset. Continually grow your knowledge of the leadership models and frameworks that guide your decisions and actions as well as those of your peers. New knowledge or the affirmation of existing knowledge can strengthen your strategic efforts and resolve.
5. *Invest time in self-care*—Pursuing a semblance of equilibrium while grappling with the constant demands of status quo resistance and the human interactions required to mobilize a support network can test the resolve of the strongest of leaders. Professional society change agents often experience diminished capacity due to the nature of these leadership roles. Fundamental to restoring leadership strength and sustaining efforts is to invest time in self-care. Determine what brings joy and replenishment and be intentional in scheduling time for that activity. Whether you choose to take a daily walk in a nearby park or in your neighborhood, just do it.

Discussion and Reflection Questions

- What systems and system components are in play within my organization that influence and impact DEI dynamics? How can I adapt, lead, and overcome challenges?
- How can I apply network theory and adaptive leadership to examine DEI issues within my organization and advance organizational change?
- How can I leverage inter- and intraorganizational connections to advance DEI reform within my organization?
- How will I develop, expand, and apply a growth mindset in my DEI change work?
- How will I take care of myself while trying to advance more just, diverse, equitable, and inclusive disciplinary and professional societies?

Conclusions

In this chapter, we shared reflections from a professional society leader who has had to navigate the complicated waters of DEI reform within her organization. We presented a conceptual framework with two components (adaptive leadership and network theory) that help frame the case study and provide insight for society leaders likewise engaging in the long yet rewarding path of change. This case demonstrates the vital need for building strong connections within and outside of the organization and centered on the importance of looking inward as well, such as developing an equity mindset and seeing clearly the operant systems in play that influence and shape DEI in societies. This case also demonstrated the need to take sufficient time for self-care due to the demanding nature of DEI reform. Thus, society leaders who are working toward DEI change must constantly balance the external and the internal, be strategic and vulnerable, care for others and themselves, and use their heart and mind to make their organizations a better place with sweeping implications for their membership.

Note

1 The American Society of Civil Engineers' DEI website can be accessed at https://www.asce.org/diversity-equity-and-inclusion.

References

Adler, P. S., and S. W. Kwon. 2002. "Social Capital: Prospects for a New Concept." *Academy of Management Review* 27 (1): 17–40. https://doi.org/10.5465/amr.2002.5922314.

Armstrong, M. A., and J. Jovanovic. 2015. "Starting at the Crossroads: Intersectional Approaches to Institutionally Supporting Underrepresented Minority Women STEM Faculty." *Journal of Women and Minorities in Science and Engineering* 21 (2): 141–57. https://doi.org/10.1615/JWomenMinorScienEng.2015011275.

ASCE (American Society of Civil Engineers). 2021. "Policy Statement 417—Justice, Equity, Diversity, and Inclusion." Accessed July 16, 2021. https://www.asce.org/advocacy/policy-statements/ps417—justice-equity-diversity-and-inclusion.

Baillie, C., A. Pawley, and D. Riley. 2012. *Engineering and Social Justice: In the University and Beyond*. Purdue University Press.

Beede, D., T. Julian, D. Langdon, G. McKittrick, B. Khan, and M. Doms. 2011. *Women in STEM: A Gender Gap to Innovation*. US Department of Commerce Economics and Statistics Administration. Accessed January 28, 2017. http://esa.doc.gov/sites/default/files/womeninstemagaptoinnovation8311.pdf.

Bilimoria, D., and X. Liang. 2012a. "Effective Practices to Increase Women's Participation, Advancement and Leadership in US Academic STEM." In *Women in STEM Careers: International Perspectives on Increasing Workforce Participation, Advancement and Leadership*, edited by L. Lord and D. Bilimoria, 146–65. Edward Elgar Publishing.

Bilimoria, D., and X. Liang. 2012b. *Gender Equity in Science and Engineering: Advancing Change in Higher Education*. Routledge, Taylor and Francis Group.

Britton, D. M. 2017. "Beyond the Chilly Climate: The Salience of Gender in Women's Academic Careers." *Gender and Society* 31 (1): 5–27. https://doi.org/10.1177/0891243216681494.

Burt, R. S. 1992. *Structural Holes: The Social Structure of Competition*. Harvard University Press.

Cadwalader, E. L., J. M. Herbers, and A. B. Popejoy. 2014. "Disproportionate Awards for Women in Disciplinary Societies." In *Gender Transformation in the Academy*, edited by C. White Berheide, M. Texler Segal, and V. Demos, 243–63. Emerald Group.

Campbell-Montalvo, R., G. Kersaint, C. A. Smith, E. Puccia, O. Sidorova, H. Cooke, H. Wao, J. P. Martin, J. Skvoretz, G. MacDonald, and R. Lee. 2022. "The Influence of Professional Engineering Organizations on Women and Underrepresented Minority Students' Fit." *Frontiers in Education* 6: 755471. https://doi.org/10.3389/feduc.2021.755471.

Eisenhart, M., E. Finkel, L. Behm, N. Lawrence, and K. Tonso. 1998. *Women's Science: Learning and Succeeding from the Margins*. University of Chicago Press.

Faulkner, W. 2007. "'Nuts and Bolts People': Gender-Troubled Engineering Identities." *Social Studies of Science* 37 (3): 331–56. https://doi.org/10.1007/978-3-319-16172-3_2.

Foor, C. E., S. E. Walden, and D. A. Trytten. 2007. "'I Wish That I Belonged More in This Whole Engineering Group:' Achieving Individual Diversity." *Journal of Engineering Education* 96 (2): 103–15. https://doi.org/10.1002/j.2168-9830.2007.tb00921.x.

Frank, K. A., and R. Xu. 2020. "Causal Inference for Social Network Analysis." In *The Oxford Handbook of Social Networks*, edited by R. Light and J. Moody, 288–310. Oxford University Press.

Granovetter, M. 1983. "The Strength of Weak Ties: A Network Theory Revisited." *Sociological Theory* 1 (1): 201–33. https://doi.org/10.2307/202051.

Hacker, S. L. 1981. "The Culture of Engineering: Women, Workplace, and Machine." *Women's Studies International Quarterly* 4 (3): 341–53. https://doi.org/10.1016/S0148-0685(81)96559-3.

Hacker, S. L. 1989. *Pleasure, Power, and Technology: Some Tales of Gender, Engineering, and the Cooperative Workplace*. Unwin Hyman.

Hall, R. M., and B. R. Sandler. 1982. *Project on the Status and Education of Women—The Classroom Climate: A Chilly One for Women?* Association of American Colleges. Accessed January 30, 2017. http://files.eric.ed.gov/fulltext/ED215628.pdf.

Heifetz, R. A., A. Grashow, and M. Linsky. 2009. *The Practice of Adaptive Leadership: Tools and Tactics for Changing Your Organization and the World*. Harvard Business Press.

Hill, L. B. 2020. "Understanding the Impact of a Multi-institutional STEM Reform Network through Key Boundary-Spanning Individuals." *The Journal of Higher Education* 91 (3): 455–82. https://doi.org/10.1080/00221546.2019.1650581.

Johri, A., and B. Olds. 2014. *Cambridge Handbook of Engineering Education Research*. Cambridge University Press.

Joshi, A., N. Pandey, and G. Han. 2009. "Bracketing Team Boundary Spanning: An Examination of Task-Based, Team-Level, and Contextual Antecedents." *Journal of Organizational Behavior* 30 (6): 731–59. https://doi.org/10.1002/job.567.

Kadushin, C. 2002. "The Motivational Foundation of Social Networks." *Social Networks* 24 (1): 77–91. https://doi.org/10.1016/S0378-8733(01)00052-1.

Kezar, A. 2011. "What Is the Best Way to Achieve Broader Reach of Improved Practices in Higher Education?" *Innovative Higher Education* 36 (4): 235–47. https://doi.org/10.1007/s10755-011-9174-z.

Krackhardt, D. 1992. "The Strength of Strong Ties: The Importance of Philos in Organizations." In *Networks and Organizations: Structure, Form, and Action, edited by Nitin Nohria and Robert G. Eccles*, 216–39. Harvard Business School Press.

Leibnitz, G. M., D. L. Gillian-Daniel, R. Greenler, R. Campbell-Montalvo, H. Metcalf, V. A. Segarra, J. W. Peters, S. Patton, A. Lucy-Putwen, and E. L. Sims. 2022a. "The Inclusive Professional Framework for Societies: Changing Mental Models to Promote Diverse, Equitable, and Inclusive STEM Systems Change." *Frontiers in Sociology* 6: 784399. https://doi.org/10.3389/fsoc.2021.784399.

Leibnitz, G. M., J. Peters, R. Campbell-Montalvo, H. Metcalf, A. Lucy Putwen, D. L. Gillian-Daniel, E. L. Sims, and V. A. Segarra. 2022b. "Refining a DEI Assessment Tool for Use in Optimizing Professional STEM Societies for Gender Equity." *Frontiers in Sociology* 7: 755372. https://doi.org/10.3389/fsoc.2022.755372.

Levinson, B. A., D. E. Foley, and D. C. Holland. 1996. *The Cultural Production of the Educated Person: Critical Ethnographies of Schooling and Local Practice*. State University of New York Press.

Lohan, M., and W. Faulkner. 2004. "Masculinities and Technologies: Some Introductory Remarks." *Men and Masculinities* 6 (4): 319–29. https://doi.org/10.1177/1097184X03260956.

McIlwee, J. S., and J. G. Robinson. 1992. *Women in Engineering: Gender, Power, and Workplace Culture*. State University of New York Press.

Pawley, A., and K. L. Tonso. 2011. "'Monsters of Unnaturalness:' Making Women Engineers' Identities via Newspaper and Magazines (1930–1970)." *Journal of Society of Women Engineers–60th Anniversary Edition* 60: 60–75.

Riley, D., A. E. Slaton, and A. L. Pawley. 2014. "Women and Minorities in Engineering." In *Cambridge Handbook of Engineering Education Research*, edited by A. Jorhi and B. Olds, 335–56. Cambridge University Press.

Rowe, M. 1974. "Saturn's Rings: A Study of the Minutiae of Sexism Which Maintain Discrimination and Inhibit Affirmative Action Results in Corporations and Non-profit Institutions." *Graduate and Professional Education of Women*: 1–9.

Rowe, M. 1975. "Saturn's Rings II, with Racist and Sexist Incidents from 1974 and 1975." *Harvard Medical Alumni Bulletin* 50 (1): 14–18.

Rowe, M. 2008. "Micro-affirmations and Micro-inequities." *Journal of the International Ombudsman Association* 1 (1): 45–8. http://web.mit.edu/ombud/publications/micro-affirm-ineq.pdf.

Segarra, V. A., C. Primus, G. A. Unguez, A. Edwards, C. Etson, S. C Flores, C. Fry, A. N Guillory, S. L Ingram, M. Lawson, R. McGee, S. Paxson, L. Phelan, K. Suggs, L. R. Vega, E. Vuong, J. C. Havran, A. Leon, M. D. Burton, J. L. Lujan, and M. Ramirez-Alvarado. 2020a. "Scientific Societies Fostering Inclusivity through Speaker Diversity in Annual Meeting Programming: A Call to Action." *Molecular Biology of the Cell* 31 (23): 2495–501. https://doi.org/10.1091/mbc.E20-06-0381.

Smith, C. A. S., H. Wao, G. Kersaint, R. Campbell-Montalvo, P. Gray-Ray, E. Puccia, J. P. Martin, R. Lee, J. Skvoretz, and G. MacDonald. 2021. "Social Capital from Professional Engineering Organizations and the Persistence of Women and Underrepresented Minority Undergraduates." *Frontiers in Sociology* 6: 671856. https://doi.org/10.3389/fsoc.2021.671856.

Solebello, N., M. Tschirhart, and J. Leiter. 2016. "The Paradox of Inclusion and Exclusion in Membership Associations." *Human Relations* 69 (2): 439–60. https://doi.org/10.1177/0018726715590166.

Tonso, K. L. 1996. "The Impact of Cultural Norms on Women." *Journal of Engineering Education* 85 (3): 217–25. https://doi.org/10.1002/j.2168-9830.1996.tb00236.x.

Tonso, K. L. 1999. "Engineering Gender—Gendering Engineering: A Cultural Model for Belonging." *Journal of Women and Minorities in Science and Engineering* 5 (4): 365–404. https://doi.org/10.1615/JWomenMinorScienEng.v5.i4.60.

Tonso, K. L. 2006. "Student Engineers and Engineering Identity: Campus Engineer Identities as Figured World." *Cultural Studies of Science Education* 2: 1–35. https://doi.org/10.1007/s11422-005-9009-2.

Tonso, K. L. 2007. *On the Outskirts of Engineering: Learning Identity, Gender, and Power via Engineering Practice.* Sense Publishers.

Tonso, K. L. 2014. "Engineering Identity." In *Cambridge Handbook of Engineering Education Research*, edited by A. Johri and B. M. Olds, 267–82. Cambridge University Press.

Trautner, J. J., K. C. Chou, J. K. Yates, and J. Stalnaker. 1996. "Women Faculty in Engineering: Changing the Academic Climate." *Journal of Engineering Education* 85 (1): 45–51. https://doi.org/10.1002/j.2168-9830.1996.tb00207.x.

Trower, C. A. 2008. "Competing on Culture: Academia's New Strategic Imperative." Presented at New Norm of Faculty Flexibility: Transforming the Culture of Science and Engineering Conference, Iowa State University, Ames, Iowa, October 11. Accessed January 30, 2017. http://www.provost.iastate.edu/sites/default/files/uploads/faculty%20resources/professional%20development/2008_10-11trower_hoc.pdf.

Tushman, M. L., and T. J. Scanlan. 1981. "Boundary Spanning Individuals: Their Role in Information Transfer and Their Antecedents." *Academy of Management Journal* 24 (2): 289–305. https://doi.org/10.5465/255842.

Walker, L. J. 1973. "Social Problems: A Concern for Civil Engineers?" *Engineering Issues: Journal of Professional Activities* 99 (1): 53–7. https://doi.org/10.1061/EGISBD.0000133.

Wulf, W. W. 2001. "Diversity in Engineering." *Leadership and Management in Engineering* 1 (4): 31–5. https://doi.org/10.1061/(ASCE)1532-6748(2001)1:4(31).

5

Raising Equity Values with the Inclusive Professional Framework and the ISO 30415: 2021 Standard for Societies (REVIIS, "Revise")

Gretalyn Leibnitz, James Felton Keith, Lisa Black, and Brian A. Burt

Disciplinary excellence is only as good as the talent base that informs it; the policies and practices that support optimal engagement by that talent base; and the processes and strategies that ensure agile, adaptive, and continuous improvement.[1] Apprehension about the adequacy of the US STEM workforce has been a congressional concern for more than sixty years (Sargent 2017). Recent anti-diversity, equity, and inclusion legislation (e.g., Texas' SB-17 [University of Texas, Austin 2024]) that bans the application of evidence-based practices to foster engagement by underrepresented populations has some fearing that US STEM global competitiveness and capacity to meet national workforce needs will be further compromised unless informed and immediate actions are taken. This pushback is a predictable result of the perceptions of injustice and marginalization of some majoritized (e.g., white, male, heterosexual, cis-gendered, and able-bodied) people. With the emergence of majoritized people's cries of discrimination, the DEI bubble (Gaudiano 2024) has burst. Now, the objective is to ensure that the baby (i.e., more than 20 years of DEI-related research) is not thrown out with the political bathwater, and that continuous improvement (based on an inclusive body of research) guides national and STEM disciplinary efforts toward an ever-evolving understanding of excellence. The focus of this chapter is to *introduce* the International Organization for Standardization's Human Resource Management Diversity and Inclusion Standard, ISO 30415: 2021 (referred to as "ISO 30415" for brevity) (ANSI 2021); *discuss* how the ISO 30415, in conjunction with tools like the inclusive professional framework for societies (IPF: Societies) (Leibnitz et al. 2022a), can be leveraged to optimize workforce engagement and resilience; and *demonstrate* in a case study how early adoption of ISO 30415 is currently being manifested in one engineering professional society.

Given "the past, current, and anticipated explosive growth in technology, and that engineering and computing account for more than 80% of the STEM workforce, engineering and computing stand out from the broader STEM category as the fields that offer the best opportunities for the greatest number of people" (Corbett and Hill 2015, 2). Yet, engineering has historically been, and continues to be, a domain in which a narrow population of people contribute, with women and people from underrecognized groups the least likely to be represented in national engineering workforces (NCSES 2017). This resulting myopic lens influences the nature of problems identified and addressed as well as undermines how efficiently and effectively problems are solved. As early as 2002, this lack in engineering was evident to Wulf (then president of the National Academy of Engineering) who made a compelling argument for the absolute necessity for diversity in engineering (Wulf 2002, 8). More recently, Gallimore (2021) expressed concern that inequities are baked in to engineering products as reflected in such examples as: women are more likely to die or be seriously harmed in car accidents because car crash dummies are modeled after men; engineering algorithms disproportionately result in keeping poor people out of jobs and housing; dark-skinned people are wrongly identified because systems are optimized on light-skinned people; and people with mobility issues have largely been left out of the design process of autonomous vehicles, which are argued to have great potential benefit for such disabled individuals. Engineers need to identify problems and offer solutions for a broad domain space.

Thanks to two decades of NSF ADVANCE-funded research, change leaders have data-informed results that help articulate the ways in which STEM systems and organizations have inadvertently set up barriers to disciplinary excellence through the exclusion of needed talent as well as evidence-based policies, practices, and strategies that offer solutions. Unfortunately, awareness and adoption of these promising practices are often siloed and localized. The 2021 release of ISO 30415 (ANSI 2021) and IPF: Societies (Leibnitz et al. 2022a) offers an exciting opportunity for standardizing disciplinary and organizational best practices. Given that engineering societies (i.e., professional and disciplinary societies) shape disciplinary culture, inform education, and produce standards (Borman et al. 2010; Chanderbhan-Forde et al. 2012), they are in a unique position to promote systemic change and are thus the focus of the REVIIS project. The goal of the REVIIS project is to increase awareness and, ideally, the application of ISO 30415 and IPF: Societies to extend and optimize engineering workforce engagement, resilience, productivity, and the acquisition and retention of talent. We, REVIIS leaders, contend that, by integrating DEI-related research results with an understanding of the arguments for DEI pushback, we can provide engineering leaders, members, and external stakeholders with an opportunity to increase awareness and knowledge as well as the resources and tools to standardize organizational change that promotes the inclusion of all engineering talent and broad appeal of products.

REVIIS work is informed by, and extends the work of, other projects referenced in this book. Specifically, REVIIS leaders benefited from Alliance to Catalyze Change for Equity in STEM Success (ACCESS) leader insights (Chapter 11), which included the importance of engagement of senior leaders early and often, common benchmarks, application of evidence-based practices, and strategies for assessment and collective improvement. The Amplifying the Alliance to Catalyze Change for Equity in STEM Success (ACCESS+) project (Chapter 12) informed tools applied in the REVIIS project. Notably, Leibnitz (ACCESS+ original principal investigator [PI] as well as REVIIS co-PI and project director)—along with Dr. Donald L. Gillian Daniel (INCLUDES Aspire Alliance PI, original ACCESS+ co-PI, and REVIIS PI) and Dr. Jan Peters (ACCESS+ consultant and REVIIS consultant)—helped adapt the INCLUDES Aspire Alliances' inclusive professional framework for societies (Leibnitz et al. 2022a), create and finalize the ACCESS+ Equity Environmental Scanning Tool (EEST) (Leibnitz et al. 2022b), and broaden the diverse disciplines represented under the ACCESS+ umbrella (e.g., aquatic science and engineering societies, including the American Society of Civil Engineers [ASCE] [Chapter 4]). The ACCESS+ EEST informed the REVIIS crosswalk that translates professional society functions to the ISO 30415; the IPF: Societies helps leaders extend their capacity for inclusive change; and the ACCESS+ cohort member ASCE provides the foundation for the REVIIS case study presented in this chapter (i.e., ASCE is one of three REVIIS co-PI engineering professional societies, including the American Society of Mechanical Engineers [ASME] and the American Society for Engineering Education [ASEE]).

Although the REVIIS project has two core foci, ISO 30415 and the IPF: Societies, because the IPF: Societies is detailed elsewhere (Leibnitz et al. 2022a) this chapter focuses primarily on ISO 30415. Although change leaders and researchers now know a lot about exclusive practices and organizational change, what has only recently emerged is an expert consensus, in the form of a global standard (i.e., ISO 30415), about what an inclusive organization looks like. Now, what is needed is the awareness of ISO 30415, a collective understanding of why inclusion promotes a competitive edge, and a standardized service management process for organizational change. In this chapter, we seek to enhance awareness, interest, readiness, and engagement with ISO 30415 and the certification process. In this case study we answer the question, "What have we learned, to date, about implementing the certification process within an engineering society?"

Background for the Case Study

As the last chapter in this part about professional societies nurturing visible thought leaders, we have the luxury of drawing on all prior work. To that end, in this background section we will briefly summarize relevant premises well-articulated by colleagues in other chapters and proceed with information specific to the REVIIS project.

- *Premise 1*—Societies are key levers for culture and system change.
- *Premise 2*—We know a lot about practices that minimize the full engagement of people from underrecognized groups.
- *Premise 3*—Inclusive change can be accomplished.

Despite the information that supports these premises, results are not unequivocally supportive of DEI efforts. The NASEM (2018a) concluded that although many societies desire to understand the experiences of professionals from underrecognized groups and incorporate strategies to address inequities, they lack critical information on how to do so. Further, arguments against DEI are that much of the work has been ad hoc and expensive with arguably little evidence of lasting change (Newkirk 2019). Although we acknowledge the limitations of prior change efforts to promote the optimal engagement of all talents, we also recognize that although not sufficient, previous work has been developmentally necessary. Rarely can one go from standing to running without falling, as one learns the nuances of balance and effort. So, honoring prior DEI thought leadership and research as well as noted criticisms of the DEI work, we proceed with integrating the two to provide background detail that informs the REVIIS approach to systemic engineering disciplinary change.

Kania et al. (2018) identified several factors that maintain system status quo and that, logically, also serve as important levers for change. Explicit factors (e.g., organizational policies, procedures, and resource flow) as well as semi-explicit factors (e.g., relationships and power dynamics) are relatively easy to change. However, these organizational components are held in place by mental models—"deeply held beliefs and assumptions, and taken-for-granted ways of operating that influence how we think, what we do, and how we talk" (Kania et al. 2018, 4) that are challenging to change because of their implicit nature. So, within professional societies, for example, Solebello et al. (2016) found that society leaders experience tension between articulated inclusion values and the drive to protect the exclusivity of the profession and the society's history and culture. Ultimately, how this tension is resolved is mirrored in society operations, for example, in ideas about what research topics deserve funding or recognition, who is qualified to hold leadership positions, or whose research or scholarship deserves to be highlighted in convenings or society publications. Because society mental models represent how things are done, they can create counter-pressure to needed change.

Gaudiano (2024) argues that counter-pressure created by mental models need not exist if those seeking change understand the "reality of those running organizations." Based on ACCESS+ work with professional society cohorts and the comprehensive EEST data collected, professional society leaders are most concerned with five areas of focus: governance and leadership, membership, convenings, recognition, and professional development (Leibnitz et al. 2022a, see the outer five petals in Figure 5.1). Specifically, professional society leaders are concerned with optimizing a competitive edge by demonstrating, and marketing, cutting-edge leadership through gold standard certifications and other means; limiting and mitigating liability; attracting and retaining staff, leaders, and members; and reducing expenses and maximizing revenue associated with deliverables like convenings, professional development for members, and publications (e.g., books and journals). From an organizational leadership perspective, all of these things are achieved through asset management.

Given that people are the greatest asset for S&P 500 market value (Ocean Tomo 2022), and arguably the same for professional societies, it follows that the management of human assets is key to optimal performance. Further, it is well acknowledged that asset diversification is also key for optimal performance. Based on these facts, Gaudiano (2024) makes a compelling argument for managing a company's human assets by optimizing and nurturing diversification by systematically eliminating factors that limit the satisfaction and engagement of employees. This orientation is informed by an understanding that the

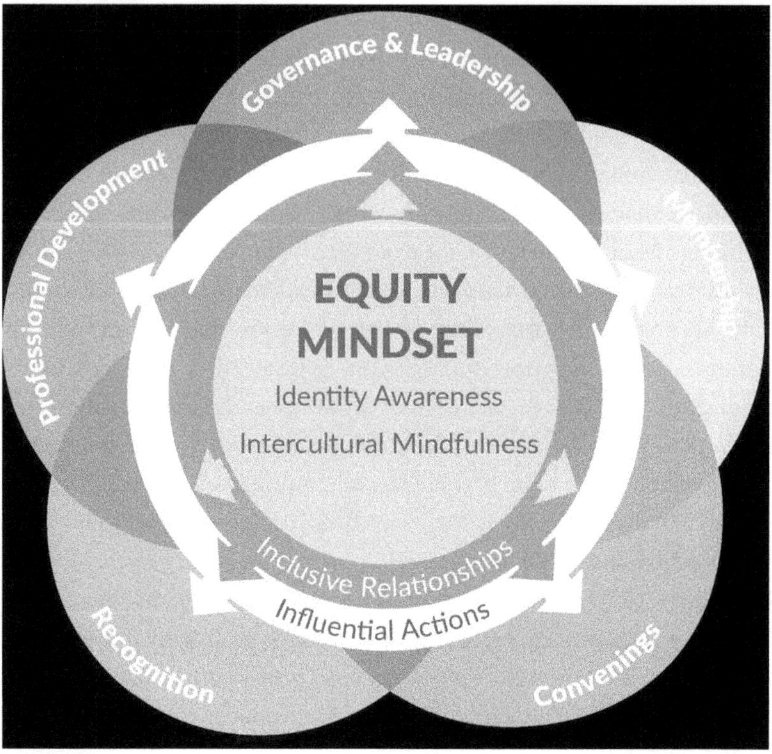

Figure 5.1 The inclusive professional framework for societies (IPF: Societies) as related to the five most common functions provided by professional societies. Reprinted from Leibnitz et al. 2022a.

company will attract and retain people, and produce products attractive to a diverse audience, the more diversified that its satisfied employees are.

Gaudiano argues that, as tracking finances is core to corporate success, leaders that pay attention to factors that minimize employee dissatisfaction and exclusion, which rob workforce productivity and foster unwanted turnover, will gain a competitive edge. Gaudiano and his Aleria team have collected "exhaustive data showing that the level of inclusion of different groups of people within an organization impacts satisfaction, which, in turn, is known to impact both productivity and attrition (retention) rates" (Aleria 2024.)

Unlike typical corporate settings, however, engineering society leaders have a unique management context in that they need to support paid staff (who may represent a small number, depending on the size of the society) and volunteers (who may significantly outnumber staff). The "reality of the staff–volunteer cooperation is akin to a double-helix DNA model with helical advancing strands of volunteers and staff structurally linked together at every twist and turn" (unpublished report by T. J. Perry, 1[2]). Leadership responsibilities and influence exist at all levels in the staff and volunteer strands, not just at the top level. Further, effective change efforts on one side can be derailed by attitudes and behavior on the other side. So, society directors need to not only be aware of both staff and volunteer experiences but also limit the liability associated with both.

Another complication is that, like most people, society leaders (who by definition are successful in the society) are least likely to be conscious and aware of factors that influence society participation by people with identities and characteristics different from their own (Gaudiano 2024). This is where the IPF: Societies can be a helpful model to guide leaders in learning what they need to know about fostering organizational inclusion. With an understanding that society leaders seek to optimize key professional domains represented in the petals of Figure 5.1, the process of fostering inclusive organizational change starts with identity awareness (both personal and organizational) and intercultural mindfulness.

Identity awareness and intercultural mindfulness, together, promote an equity mindset. An equity mindset is critical for optimizing organizational engagement. Actively establishing inclusive relationships with a variety of people and organizations (within and outside the organization) as well as collecting satisfaction data and information regarding barriers to organizational engagement helps leaders discern where and what actions are needed to reduce dissatisfaction and promote optimal engagement. For more information about the IPF: Societies, please review work by Leibnitz et al. (2022a). Ultimately, even with guidance and tools for fostering the inclusion of staff and volunteers, society leaders mostly engage in ad hoc activities. Now, with the ratification of the ISO 30415, a map to standardize human asset engagement and diversification throughout the organization is available.

The value and development of standards are championed by the International Organization for Standardization (ISO). This nonprofit worldwide federation of 163 national standards bodies (e.g., the American National Standards Institute is a national standards body for the United States) has developed and published over 22,000 standards "of virtually every possible sort" that range from standards on information technology to human resource management. The existence of ISO is founded on the belief that "although largely invisible in our daily lives, [international standards] are a crucial component to making things safer and better in the world around us" (ISO 2023). The ISO defines a standard as the expert consensus on evidence-based practices that determine what "great looks like" and promotes consistency in everyday products and services to "ensure safety, reliability, and quality [through] sustainable, ethical practices" (ISO 2024a).

In a global economy standards are important because they create a common language and process for producing goods and services. Grounded in evidence, standards ensure rigor and provide benchmarks for excellence. Research demonstrates that standards like the Quality Management Standard (ISO 9001: 1978) confer significant benefit to companies and organizations (Neyestani and Juanzon 2017). Through ongoing iteration, standards provide a means for collective continuous improvement at local and global levels and a means of reducing liability. Consequently, it follows that compliance ensures a global competitive edge and, conversely, non-compliance undermines a global competitive advantage.

"Like a symphony, it takes a lot of people working together to develop a standard. The ISO's role is similar to that of a conductor, while the orchestra is made up of independent technical experts nominated by ISO members. Independent experts begin the process with the development of a draft that meets a market need within a specific area" (ISO 2024b). The draft is then shared for commenting and further discussion. The voting process is the key to consensus. Once consensus is achieved then the draft is on its way to becoming an ISO standard. If an agreement is not reached, then the draft will be modified further and voted on again. From the first proposal to final publication, developing a standard usually takes about three years.

Specifically, the ISO 30415 provides guidelines to organizations on diversity and inclusion principles, practices, approaches, methods, and mechanisms to enable and support equity, fairness, equality, and accessibility in workplace contexts through external partners and sponsors as well as, ultimately, in deliverables. In the words of the ISO, the 30415: 2021 standard:

> provides guidance on D&I [diversity and inclusion] for organizations, including their governance body, leaders, workforce and recognized representatives, and other stakeholders. It is intended to be scalable to the needs of all types of organizations in different sectors, whether in public, private, government or non-governmental organizations, regardless of size, type, activity, industry or sector, growth phase, external influences and country-specific requirements.
>
> (ISO 2021)

Essentially, the ISO 30415 defines, through a series of statements, *what* an organization looks like when it has optimized its human capital throughout all aspects of the organization.

ISO 30415 is organized around a plan-do-check-review process. This process focuses on readiness conditions, accountability, actions, and measures and outcomes, as reflected in a series of statements.

1. *Fundamental prerequisites for diversity and inclusion*—Establishes the essential conditions necessary for effective diversity and inclusion initiatives. An example statement is: "Valuing all people intrinsically, individually, and as groups; appreciating how different diversity dimensions intersect; and acknowledging that demographic and other personal characteristics can possibly be protected by law and regulation."
2. *Associated accountabilities and responsibilities*—Clearly defines the roles and responsibilities for implementing diversity and inclusion. An example statement is: "Holding senior leaders to account when evaluating the organization's D&I [diversity and inclusion] opportunities and risks and reviewing performance and progress in achieving D&I objectives and the impact of their outcomes."
3. *Recommended actions*—Suggests specific actions to foster an inclusive environment. An example statement is: "Review plans, systems, technologies, policies, processes, and practices to ensure that they are fair, equitable, and do not adversely impact people, including those related to the use of technologies and automated systems."
4. *Suggested measures and potential outcomes*—Provides metrics and outcomes to assess and guide diversity and inclusion efforts. An example statement is: "Collect data on workforce and other stakeholder perceptions about the organization's commitment to D&I [diversity and inclusion] such as from focus groups, surveys, interviews, and external reviews."

Through this comprehensive approach, decision-makers tailor their organizational engagement based on their specific context and the challenges identified through the process of working through the standard.

The importance of the ratification of the ISO 30415 includes the rationale for all standards and is especially valuable because it marks the adoption of the first global consensus for organizational diversity and inclusion. The ISO 30415 comprehensively defines evolving inclusive organizational excellence, promotes a systematic approach, fosters accountability for continual improvement in support of human assets, can inform inclusive disciplinary accreditation and certification, and prevents risks and promotes mitigation of potential lawsuits. So, rather than the typical siloed and often ad hoc process that engineering societies historically deploy to adapt to an increasingly complex and changing global context, ISO 30415 offers a lens through which organizational structure and service delivery are ordered and optimized to a global benchmark.

ISO 30415 is the culmination of forty years of previous work and has foundations in prior standards, such as ISO 9001: 1978 (the 9000 is a family of standards that depicts a set of five quality management systems) and ISO 30414: 2018 (guidelines for internal and external human capital reporting). Although the standards demonstrated benefit to companies, Neyestani and Juanzon (2017), for example, were the first to study how specifically ISO 9001 improved organizational performance from a variety of sectors. Based on an analysis of publications from 2002 to 2017, the authors concluded that ISO 9001 positively "significantly affected internal business processes" (2017, 6). Additionally, based on an analysis of a matched sample of almost 1,000 California-based companies, Levine and Toffel (2010, 1) found that companies adopting ISO 9001 had "higher growth rates for sales, employment, payroll and average annual earnings, and injury rates declined slightly." They concluded that the "process of ISO 9001 certification appears to be valuable for most adopters" (2010, 26), "small enterprises are particularly likely to benefit from adoption" (2010, 28), and that "substantial improvement in outcomes after ISO 9001 adoption certification has lessons for organizational scholars" (Levine and Toffel 2010, 28). What ISO 30415 adds is a focus for infusing diversity and inclusion into all aspects of an organization to maximize competitive advantages and impact.

From a professional society perspective, standards are well known to many STEM societies and especially to engineering societies. Many engineering societies are technical, safety, and educational standard-setting organizations at the national and international levels. For example, engineering professional societies play an instrumental role in the creation, evolution, and assessment of degree program-accreditation standards by participating in the Accreditation Board for Engineering and Technology (ABET). ABET serves the interests of thirty-five member societies (such as ASCE, ASME, and ASEE) and represents more than 1.5 million professionals around the world. These member societies set policy, develop strategy, and conduct accreditation activities worldwide on behalf of their professions. This is done collaboratively and collectively through the ABET accreditation process and has a direct impact on over 4,674 degree programs at 920 institutions in 42 countries. In 2021 deans of engineering from Big 10+ encouraged ABET to incorporate DEI into the General Criteria for Accrediting Engineering Programs, which was approved by the ABET Board of Directors on March 27, 2022. In so doing, ABET formally committed to engineering professionalism as "incorporating the highest standard of professional integrity, dignity, fairness, justice, and respect for everyone" (ABET 2021). So, the application of and certification to ISO 30415 for engineering societies could help meaningfully inform other standards like ABET accreditation.

Further, given key society leadership pain points, like declining memberships (Walker and Nolen 2023), certification to ISO 30415 could promote membership appeal by signaling society relevance to a diverse, modern audience; validate professional society actions taken to professionally and publicly promote inclusiveness; inform broader definitions of engineering professionalism; and expand domains of disciplinary reach, thus promoting broad disciplinary excellence. Last but not least, certification to ISO 30415 could reduce liability by decreasing uninformed service, product production, and delivery.

The reduction of liability, especially in professional societies composed not only of staff but volunteer workers, is a particularly attractive aspect of ISO 30415 certification for engineering societies. Although there is little or no actual legislation that requires DEI, DEI efforts were a natural outgrowth of Title VII of the Civil Rights Act of 1964 (Pub. L. No. 88-352, 78 Stat. 241) that prohibits discrimination on the basis of race, color, religion, national origin, and gender (including sexual harassment, sexual orientation, and gender identity). Related liability laws include the Age Discrimination in Employment Act of 1967 (Pub. L. No. 90-202, 81 Stat. 602), the Americans with Disabilities Act of 1990 (Pub. L. No. 101-336, 104 Stat. 327), and the Genetic Information Nondiscrimination Act of 2008 (Pub. L. No. 110-233, 122 Stat. 881). Since the DEI concept arose organically—with each company and institution developing its own approach and often with personnel that had no training or experience in DEI—there was no uniformity of either requirements or expertise, which resulted in costly liability for violations of Title VII. We estimate a minimum of $10 billion is paid out annually by insurance companies for director and officers claims and a growing number of employment practices liability claims (NMSC 2024;

VMR 2024). Further, these lawsuits are increasing every year. The US Equal Employment Opportunity Commission reported a 30 percent increase in 2023 from 2022 workforce discrimination charges (US EEOC 2024), not including out-of-court settlements and insurance claims. The ISO 30415, through the certification process, provides a consistent path to reaching inclusion goals and reducing the ever-increasing liability risks associated with the management of human assets.

How does an organization become ISO certified? The ISO does not certify standards it develops and produces; external companies do so via a process of peer review. In the service of facilitating ISO 30415 certification, the Keith Institute (a Michigan-based not-for-profit) gave a grant to create the Diversity & Inclusion Service Management (ISO 30415: DISM) Forum. The forum acts to critique changes to the standard and deliver training on implementation of the standard. In 2021 ISO 30415: DISM established the first ISO 30415 certifying body (ISI n.d.).

Further, ISO standards typically evolve in pairs: a general standard that defines the concept, namely, *what* the standard is (e.g., ISO 30415), and a technical standard reflecting *how* the work is accomplished. As of this publication, an ISO *technical* standard for ISO 30415 does not exist. The ISO 30415: DISM Forum commissioned the Inclusion Score (www.inclusionscore.ai) to create the "Worlds #1 D&I Operating System" and standardize a DEI workflow, as informed by ISO 30415, that could also be used to inform the certification process.

Care was taken to map ISO 30415 to the Inclusion Score platform so that no matter where the organization is in its inclusion journey, the program can guide it along a path to successful ISO 30415 certification. The optimization of a company's posture is achieved by the following three key steps, the outputs of which help executives prioritize issues and effectively budget organizational improvements over time.

1. Measure the organization's diversity and inclusion policies and practices against ISO 30415.
2. Map the various domains of ISO 30415 to a four-level maturity scale that serves as a referential benchmark and improvement roadmap.
3. Generate five areas of focus for organizational improvement.

The Inclusion Score platform leverages a four-point maturity scale (Table 5.1) based on the Carnegie Mellon Capability Maturity Model (CMM)

Table 5.1 The Inclusion Score's Four-point Maturity Scale

Level	Label	Description
1	Ad hoc	Unpredictable: Some inclusive practices are taken, but there is minimal measurement, methodology, and strategy. Action is reactive.
2	Learning	Managed, but reactive: The company is beginning to structure and organize around evidence-based, inclusive strategy. Action is still reactive.
3	Under control	Defined and controlled: The company manages, processes, and implements robust controls. Action is proactive.
4	Measured for continuous improvement	Measured and optimizing: The company maintains a strategic advantage over competitors with controlled, measured, and continuously improved diversity and inclusion processes. Actions are being optimized for the future.

Sources: ISI (n.d.); adapted from Paulk et al. (1993).

(Paulk et al. 1993), to determine an organization's liability risk and optimize its human capital.

By calculating a maturity score based on the ISO 30415 standard, inequities in organizational systems, policies, processes, and practices as well as people's conscious and unconscious behaviors, can be identified and addressed. The Inclusion Score operating system is important because it (1) guides a user through the elements of ISO 30415, which is likely to greatly decrease or eliminate Title VII violations; (2) provides a means of organizing and centralizing evidence of organizational policies and practices; (3) pinpoints areas of weakness for organizational leaders to promote continuous improvement; (4) creates a means for peer comparison; and (5) fosters a solid foundation for ISO 30415 certification—all of which create a mechanism for organization leaders to gain a competitive edge over peers in an aggressive global economy.

Case Study Narrative

This REVIIS case study reflects work done for and by the ASCE. Given that this case study depicts the process to introduce and engage ASCE leaders and members to the ISO 30415: 2021 standard and the certification process, the information presented does not meet the definition of "research," so no quantitative or qualitative data has been collected requiring institutional review board approval. As noted earlier, ASCE is a REVIIS co-PI, as represented by Dr. Lisa Black (ASCE's director of cultural belonging and social ethos); and REVIIS work is informed by and extends the work of other ASCE work (e.g., ACCESS+ [Chapter 12]; and lessons from the field [Chapter 4]).

First, a crosswalk was developed to map typical functions, or services, of an engineering society to the relevant domains of ISO 30415. ACCESS+'s EEST provided an efficient, and evidence-based means of mapping the functions of a typical professional society to relevant areas of ISO 30415 (the most common five functions represented in the petals of Figure 5.1). To create the crosswalk, the REVIIS project team translated these functions into the four ISO 30415 focus areas (i.e., governance and leadership, employment, external suppliers and stakeholders, and products), the majority of which fall within ISO 30415 products (see Figure 5.2).

Second, we engaged ASCE's senior leadership team (SLT) for support and engagement. To prepare for the SLT meeting, Dr. Black provided members with a recently recorded video providing an overview of ISO 30415, a training entitled an "ISO 30415 Briefer" (Leibnitz and Keith 2024); and narrative and bulleted

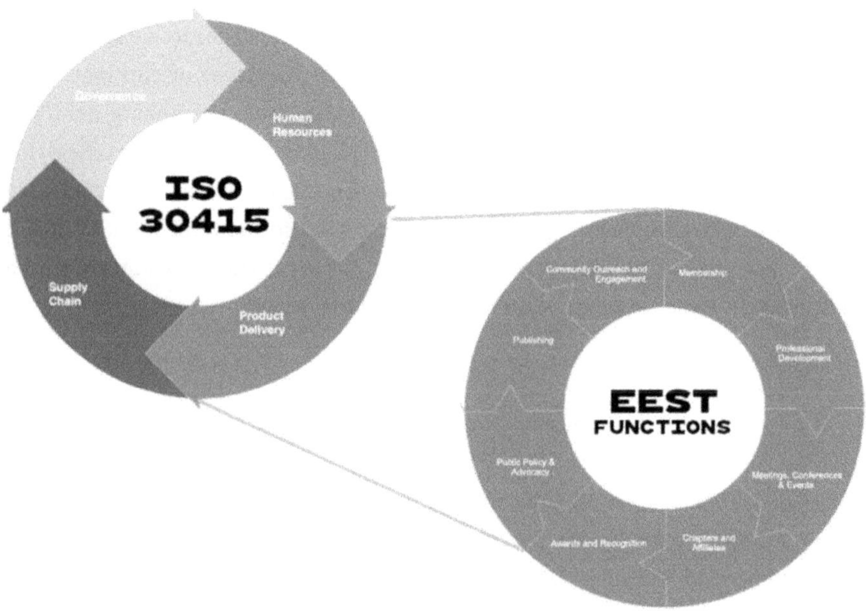

Figure 5.2 The REVIIS crosswalk. Mapping the EEST typical functions of societies to the ISO 30415. Created in Keynote by James Felton Keith.

versions of the business case for ISO 30415: 2021 certification (REVIIS 2024). At the ASCE SLT meeting the REVIIS project team was introduced, an overview of the REVIIS project was shared, the bulleted business case for ISO 30415: 2021 certification was reviewed, questions about the ISO 30415: 2021 were answered, and information was provided to clarify how the SLT could support the REVIIS work, and specifically Dr. Black's efforts, by confirming and supporting:

1. ISO 30415 certification as a priority for the organization.
2. ISO 30415 certification as a priority for ASCE stakeholder engagement (e.g., individual staff, leaders, and members).
3. Professional development for ASCE leaders and selected members to be certified as ISO 30415: 2021 auditors, at no cost, in a forthcoming training at the annual DISM Forum.
4. Professional development for ASCE leaders and members, at no cost, to receive an ISO 30415 Briefer training (i.e., a one-hour overview of ISO 30415).
5. Professional development for ASCE leaders and members interested in certification as ISO 30415 auditors (as a collaborative means of revenue generation) by offering an eight-hour, virtual, ISO 30415 certification course led by Inclusion Score representatives.

The outcome of the SLT meeting was enthusiastic support with key members of the SLT (including the president-elect) opting to participate in an in-person professional development opportunity to earn ISO 30415 auditor certification, which was offered at the 2024 DISM Forum.

Third, ASCE prepared to compile a service catalog representing all diversity and inclusion efforts and documentation. Dr. Black engaged leaders, previously identified, for the ISO 30415 certification process (see the discussion). The REVIIS project team proposed a process for collecting and onboarding information to the Inclusion Score platform. For an efficient process, the support of an ISO 30415 certified auditor was deemed necessary. Although Dr. Black and other ASCE leaders were newly certified ISO 30415 auditors, it was clear that the REVIIS team would need to provide auditor guidance. REVIIS guidance for auditors included the following actions:

- Review the society website to glean evidence of mission, policy, practices, and programs that support inclusive engagement.
- Create an Inclusion Score account for the society lead (as well as the auditor).

- Pre-populate the Inclusion Score account with society website–gleaned information.
- Launch the work in a meeting with the lead to discuss pre-populated information and assign the next steps.

- Guide the lead through review and upload of evidentiary documents for the nine domains pertinent to the professional society.
  - Initially, focus on onboarding and induction to ascertain how new members are welcomed into the organization and what new members can expect of the organization. (Arguably, they could start with the domains of governance or leadership, but the fact that auditing is a discovery process and there is a society lead engaged in the certification process, the auditor can add that information later as needed.)
  - Recommended follow up focus on inclusive culture to discover affinity groups supported by the organization. These groups may serve the lead in an advisory role to identify where exclusionary practices and dissatisfaction are occurring in the organization and provide formative feedback on actions taken to remediate concerns.
- Review the final Inclusion Score data visualizations and report with the lead.
- Discuss with the lead any next steps for the SLT to take, including submission for a peer-reviewed certification (good for two years, once certified) and a proposal for ongoing improvement.
- Support submission for certification.

Findings

Although we are early in the process of supporting ISO 30415 certification for engineering professional societies, we have made important in-roads. Specifically, we have:

- Created a crosswalk for mapping the typical functions of professional societies onto the risk management domains of ISO 30415 through the Inclusion Score platform.
- Developed a flexible model to engage and support whomever from the society has interest in the certification process, regardless of their formal organizational role.

- Developed a strong, professional society, business case for ISO 30415 certification as a resource for fostering SLT support and engagement.
- Corroborated the value of achieving early SLT support for the certification process as well as recommended consistent and regular updates to maintain SLT support and engagement.
- Articulated the need of building ISO 30415 fluency by exposing core staff and volunteer leaders to an ISO 30415 Briefer, as an important lever in supporting ISO 30415 certification and ongoing improvement practices.
- Identified the importance of having members and leaders as ISO 30415 certified auditors to help inform the certification work and sustain the process of continuous improvement.

Discussion and Reflection Questions

Conceivably, organizational change can be initiated by anyone in a professional society (e.g., student, volunteer member, staff, a person with formal positional authority within the organization, an outside stakeholder). If you are seeking to initiate the process for organizational ISO 30415 certification, it is important to reflect on your identity and connection to formal, and informal, authority within the organization. The IPF: Societies (Leibnitz et al. 2022a) may be useful to help you develop an engagement strategy. Questions to consider are:

- What positional power do you have in the organization?
- What is your sphere of influence?
- How can you leverage your positionality and influence to garner support for the ISO 30415: 2021 certification process?
- Where might you be unaware or uninformed due to the limitations of your experiences within the organization?
- Are you the right person to lead the certification effort, or might the goal of certification be better served through someone else in the organization?

Leadership support is absolutely necessary, but not sufficient, for the certification process. Engage the SLT early and as often as needed to ensure active interest, optimize vocal support for the work, and gain continued authorization for broad engagement and resource use. Reflect on both the narrative and bulleted versions of the REVIIS business case for ISO 30415: 2021 certification.

- Are there other contextual factors specific to your organization that need to be included?
- Before meeting with the SLT consider, "What might SLT member questions be?" and "What are you asking for?"
- What information presented in this chapter case study might be pertinent to your context, and how might you need to change this information?
- Review the organization chart and reflect on who you think would be useful to engage in the team supporting the certification effort. Ask SLT members, "Who would you recommend from the organization to be engaged in the certification process?"

Once you have secured leadership support, activating a core team is critical. The more diverse and inclusive the team is the greater the potential for broad and accurate information provided for the certification process. Remember, the information gathered by the people included in the team informs the documentation collected and the corresponding CMM score that is valuable to the certification process. The ISO 30415 domain CMM scores will be the basis upon which society leaders make decisions to support organizational, continuous improvement efforts, so it is critical that you consider the composition of the certification process team. Influential stakeholders and questions to consider include:

- *Leaders*—Awareness, knowledge, and active support by organizational leaders in formal positional authority is necessary, but not sufficient, for efficient and effective engagement in any adaptive change measure with the lead.
  - Key leaders on the president's board.
  - The executive director.

- *Data management specialists*—It is likely that the organization has data on its members as well as assessments providing information about organizational values and processes. This type of information should be identified and collected as a foundation for the certification process.
  - Who collects organizational data, especially about demographics on members and leaders, and are there previous DEI assessments available that you might include?
  - Where is that information collected, and what are the organizational policies regarding the release and sharing of this information?

- *Communication and change management channels*—To foster buy-in and promote efficient and effective change, a change leader needs to identify core communication channels and engage these communication channels regularly.
  - Consider what marketing and communications channels are available to you. Organize from the easiest to most difficult for you to engage (e.g., website content, newsletter, publications, listserv, LinkedIn).
- *Affinity groups*—People from organizational affinity groups can serve as key advisers in the change process because they know where some organizational barriers are, what actions are desirable, and can inform leadership decisions for prioritizing actions and resources.
  - What affinity groups does your organization have?
  - How can you engage the leaders of this group?
- *Funding partners*—People who have access to funding and other valued resources can be helpful in securing immediate funds to support the work and build capacity to sustain the work.
  - Who is involved in fundraising, gifts, and foundation work?
  - Does the organization have a grant writing lead?
  - Are there corporate or government stakeholders who may have access to external funding to support this work?
- *Volunteer coordinator*—Professional societies are unique in their reliance on volunteers; thus, it is critical to ensure that the same level of awareness building and professional development associated with the certification process be available to volunteers as to paid staff.
  - Who recruits volunteers, and how are they recruited?
  - What are the key volunteer committees, and how are leaders of these committees selected?
  - Who onboards volunteer leaders, and what type of policies, practices, and materials are used to onboard volunteers?
- *Standards committee*—Engineering professional societies often create, as well as seek to uphold, standards in product development and quality management, so representation from this group will be valuable to the ISO 30415 certification process.
  - Who leads the standard committee?
  - What is the typical process for promoting standard awareness and, where appropriate, the decision to move toward certification?

- *Professional development*—The life blood of a professional society is arguably professional development, so it is important that the society offer ISO 30415: 2021 standard information and possible auditor certification for their members to be efficiently and effectively engaged in supporting and maintaining disciplinary excellence.
  - Who is involved in offering professional development for members, in general?
  - Is there someone that offers specialized professional development to members and leaders?
  - Can we leverage REVIIS to support ISO 30415 certification professional development opportunities to members in a way that also builds revenue for organizational ISO 30415 compliance funding? For example, by partnering with Inclusion Score, Inc., with some percentage of the funding going to support your DEI efforts.
- *Society-specific circumstances*—Identify other key influencers in your organization.
  - Consider accreditation partners (e.g., ABET or other regional accreditors), policy leaders, chapter leaders, etc.

It is very helpful to have the lead, or at least a member of the team, be a certified ISO 30415: 2021 auditor. Who is best poised to become a certified auditor (e.g., familiar with ISO standards, DEI research, and organizational change practices; have a broad formal and informal circle of influence; leadership experience with the organization; experience with accreditation processes and software)? Because it is also helpful to have team and organizational members familiar with the ISO 30415 language and process, what are ways that you can deliver ISO 30415 Briefer information broadly and consistently?

Finally, it is important that the lead for the certification process serve as a boundary spanner between internal and external stakeholders. As this person will serve to communicate volunteer and employee perceptions and dissatisfactions to organizational leaders, they must be trusted, especially by affinity groups.

Conclusions

In this chapter, we outline rationale for why and how standards promote systemic change. We specifically highlight the ISO 30415 as a means of making engineering and engineers better able to address and meet global needs. Now, more than any time in history, the maximization of broad leadership, employee, and volunteer talent as well as optimal engagement is critical for ongoing engineering organizational and disciplinary success. Global standards, such as ISO 30415: 2021, provide certified companies and their leaders with a competitive edge through the exemplary management of human capital. Although the backlash against DEI is resulting in changes to practices, the outcome (when paired with awareness and evidence-based knowledge gained through research) can result in an even greater workforce resilience, the attraction and retention of broad talent, and a reduction in lost productivity and liability. We believe that projects like REVIIS offer vehicles to foster adaptive and agile leaders capable of meeting the need for ever-evolving engineering excellence.

Next steps for REVIIS project leaders are many and varied. We will continue to, and invite help in, building awareness about the ISO 30415 and resources like the IPF: Societies, the ISO 30415: DISM Forum, and the Inclusion Score platform to support organizational inclusion. We will refine continuous improvement strategies that help organizational leaders reduce barriers to the full engagement of all talent, as reflected in all the ways that people manifest. Finally, we will continue to initiate and support ongoing collaboration with colleagues whose work is represented in these volumes, accrediting agencies such as ABET, institutions of higher education, government sectors such as the US Department of Defense, and all who seek the highest engagement in engineering and other STEM disciplines. The world needs all of our talent.

Notes

1 The corresponding author (Gretalyn Leibnitz) would like to honor and recognize the early contributions of Donald L. Gillian-Daniel (the original REVIIS PI) and Robin Greenler, without whom there would be no REVIIS project. Jan Peters (REVIIS consultant and chief executive officer of Katalytik) has been a stalwart archaeologist in excavating and making clear the process of organizational change. Tom Perry, Dawn Bennett-Alexander, Effenus Henderson, Karl Reid, and Ala Qubbaj (members of the REVIIS external advisory board), I celebrate for their contributions to this chapter, their support of the REVIIS project, and their tenacious leadership and commitment to inclusive organizations.

Deep appreciation must also be extended to the many colleagues that have contributed to my understanding and organizational change expertise over the years, including many contributors to this volume (e.g., Veronica Segarra, Lucas Hill, Heather Metcalf, and Ershela Sims). The process of change is made easier, and the work is made better, together.

2 This resource is expected to become available on the REVIIS website under the title "The Business Case for the Everyday Professional Society CEO Regarding Diversity and Inclusion and the Role of the D&I [Diversity and Inclusion] Standard (ISO 30415: 2021)."

References

ABET (Accreditation Board for Engineering and Technology). 2021. "Inclusion, Diversity, Equity, & Accessibility." https://www.abet.org/about-abet/idea-inclusion-diversity-equity-accessibility/.

Aleria. 2024. "Inclusion Impact Calculator." https://www.aleria.tech/inclusion-calculator.

ANSI (American National Standards Institute). 2021. "New ISO Standard Supports Diversity and Inclusion in the Workplace." Press release May 4. https://www.ansi.org/standards-news/all-news/2021/05/5-4-21-new-iso-standard-supports-diversity-and-inclusion-in-the-workplace.

Borman, K., R. Halperin, and W. Tyson, eds. 2010. *Becoming an Engineer in Public Universities: Pathways for Women and Minorities*. Springer.

Chanderbhan-Forde, S., R. Heppner, and K. Borman. 2012. "'The Doors Are Open' but They Don't Come In: Cultural Capital and the Pathway to Engineering Degrees for Women." *Journal of Women and Minorities in Science and Engineering* 18 (2): 179–98. https://doi.org/10.1615/JWomenMinorScienEng.2012004190.

Corbett, C., and C. Hill. 2015. *Solving the Equation: The Variables for Women's Success in Engineering and Computing*. American Association of University Women. https://www.aauw.org/app/uploads/2020/03/Solving-the-Equation-report-nsa.pdf

Gallimore, A. D. 2021. "It's Time for Engineering to Be Equity-Centered." *Inside Higher Ed*, August 29. https://www.insidehighered.com/views/2021/08/30/diversity-equity-and-inclusion-should-be-required-engineering-schools-curricula#.

Gaudiano, P. 2024. *Measuring Inclusion: Higher Profits and Happier People, without Guesswork or Backlash*. Practical Inspiration Publishing.

ISI (Inclusion Score, Inc.). n.d. "ISO 30415: 2021 Certification." Accessed December 1, 2024. https://www.inclusionscore.ai/certification.

ISO (International Organization for Standardization). 2021. "ISO 30415: 2021(en) Human Resource Management – Diversity and Inclusion." https://www.iso.org/obp/ui/#iso:std:iso:30415:ed-1:v1:en.

ISO (International Organization for Standardization). 2023. *ISO Strategy 2030*. ISO. https://www.iso.org/files/live/sites/isoorg/files/store/en/PUB100364.pdf.

ISO (International Organization for Standardization). 2024a. "International Organization for Standardization Home Page." https://www.iso.org/home.html.

ISO (International Organization for Standardization). 2024b. "Developing Standards." https://www.iso.org/developing-standards.html.

Kania, J., M. Kramer, and P. Senge. 2018. *The Water of Systems Change*. FSG. https://www.fsg.org/wp-content/uploads/2021/08/The-Water-of-Systems-Change_rc.pdf.

Leibnitz, G., and J. F. Keith. 2024. "Leveraging the Diversity and Inclusion Standard (ISO 30415: 2021) for Systemic Change." ADVANCE Implementation Mentors Network ISO 30415: DISM Briefer, August 27. YouTube. Video, 38 min., 51 sec. https://www.youtube.com/watch?v=PU1BOW4Qnc8.

Leibnitz, G. M., D. L. Gillian-Daniel, R. Greenler, R. Campbell-Montalvo, H. Metcalf, V. A. Segarra, J. W. Peters, S. Patton, A. Lucy-Putwen, and E. L. Sims. 2022a. "The Inclusive Professional Framework for Societies: Changing Mental Models to Promote Diverse, Equitable, and Inclusive STEM Systems Change." *Frontiers in Sociology* 6: 784399. https://doi.org/10.3389/fsoc.2021.784399.

Leibnitz, G. M., J. Peters, R. Campbell-Montalvo, H. Metcalf, A. Lucy Putwen, D. L. Gillian-Daniel, E. L. Sims, and V. A. Segarra. 2022b. "Refining a DEI Assessment Tool for Use in Optimizing Professional STEM Societies for Gender Equity." *Frontiers in Sociology* 7: 755372. https://doi.org/10.3389/fsoc.2022.755372.

Levine, D. I., and M. W. Toffel. 2010. "Quality Management and Job Quality: How the ISO 9001 Standard for Quality Management Systems Affects Employees and Employers." Working Paper No. 09-018. Harvard Business School, January 18.

NASEM (National Academies of Sciences, Engineering, and Medicine). 2018a. *Engineering Societies' Activities in Promoting Diversity and Inclusion: Proceedings of a Workshop—in Brief*. The National Academies Press. https://doi.org/10.17226/25323.

NCSES (National Center for Science and Engineering Statistics). 2017. *Women, Minorities, and Persons with Disabilities in Science and Engineering: 2017*. Special Report NSF 17-310. National Science Foundation. https://wayback.archive-it.org/5902/20240719203439/https://www.nsf.gov/statistics/2017/nsf17310/citation.cfm.

Newkirk, P. 2019. "The Diversity Business Is Booming, but What Are the Results?" *Time*, October 10. https://time.com/5696943/diversity-business/.

Neyestani, B., and J. B. P. Juanzon. 2017. "ISO 9001 Standard and Organization's Performance: A Literature Review." *International Journal of Advanced Multidisciplinary Research* 4 (2): 6–13. http://dx.doi.org/10.22192/ijamr.2017.04.02.002.

NMSC (Next Move Strategy Consulting). 2024. *Directors and Officers (D&O) Liability Insurance Market by Coverage (Employment Practice Litigations, Regulatory Investigations, Customer Suits, Accounting Irregularities and Other Stakeholder Claims), and by End User (Public Organizations, Private Organizations, and Non-Profit Organizations)-Global Opportunity Analysis and Industry Forecast, 2024–2030.*

https://www.nextmsc.com/report/directors-and-officers-dando-insurance-market#:~:text=Market%20Overview,9.90%25%20from%202024%20to%202030.

Ocean Tomo. 2022. *Ocean Tomo Intangible Asset Market Value Study*. J. S. Held. https://www2.oceantomo.com/l/710293/2022-03-29/6dwh16/710293/1648600589q2zQV1QZ/Ocean_Tomo_IAMV_Study_Report_03_28_22.pdf.

Paulk, M. D., B. Curtis, M. B. Chrissis, and C. V. Weber. 1993. *Capability Maturity Model for Software, Version 1.1*. Technical Report CMU/SEI-93-TR-024 ESC-TR-93-177. Software Engineering Institute, Carnegie Mellon University. https://insights.sei.cmu.edu/documents/1092/1993_005_001_16211.pdf.

REVIIS. 2024. *REVIIS ADVANCE Partnership: Raising Equity Values with the Inclusive Professional Framework and ISO DEI Standards in Societies (REVIIS)*. https://www.inclusionscore.ai/reviis.

Sargent, J. F. 2017. *The U.S. Science and Engineering Workforce: Recent, Current, and Projected Employment, Wages, and Unemployment*. Congressional Research Service Report No. 7-5700. https://sgp.fas.org/crs/misc/R43061.pdf.

Solebello, N., M. Tschirhart, and J. Leiter. 2016. "The Paradox of Inclusion and Exclusion in Membership Associations." *Human Relations* 69 (2): 439–60. https://doi.org/10.1177/0018726715590166.

University of Texas, Austin. 2024. "Texas Senate Bill 17—Prohibition of Diversity, Equity, and Inclusion Activities." *University Risk and Compliance Services*, August 12. https://compliance.utexas.edu/sb17.

US EEOC (United States Equal Employment Opportunity Commission). 2024. "EEOC Releases Annual Performance Report for Fiscal Year 2023." Press release March 11. https://www.eeoc.gov/newsroom/eeoc-releases-annual-performance-report-fiscal-year-2023.

VMR (Verified Market Research). 2024. *Global Employment Practices Liability Insurance Market Size by Coverage Type, by End-User, by Distribution Channel, by Geographic Scope and Forecast*. VMR, September. https://www.verifiedmarketresearch.com/product/employment-practices-liability-insurance-market/.

Walker, A. L., and L. Nolen. 2023. "Bowling Alone and Leaving Students Behind: Placing ASEE Civil Engineering Division Membership Trends in Context." Paper presented at 2023 American Society for Engineering Education Annual Conference and Exposition. Baltimore, Maryland, June. Paper No. 37393. https://sftp.asee.org/bowling-alone-and-leaving-students-behind-placing-asee-civil-engineering-division-membership-trends-in-context.pdf.

Wulf, W. A. 2002. "The Importance of Diversity in Engineering." In *Diversity in Engineering: Managing the Workforce of the Future*, edited by C. Reed, S. Begay-Campbell, S. Brainard, D. Chubin, J. B. Cruz, Jr., W. Friend, S. Jenniches, C. Lasser, B. Montoya, D. M. Porter, Jr., and J. West, 8–14. National Academies Press. https://doi.org/10.17226/10377.

Part Two

Organizations Dedicated to Inclusive Infrastructure

Introduction to Part II

Scientific societies can help the disciplines they represent create infrastructures where inclusivity can flourish. In doing so, societies foster disciplinary spaces that can curb the impact of social forces that often prevent inclusive practices from taking root, for example, implicit bias and overvaluing the technical aspects of science.

The five chapters in this part share and discuss case studies that help us reflect on the impact of inclusive practices in disciplinary communications by society journals, the inclusion of minority-serving and emerging research institutions into the research landscape, the incorporation of inclusive values into long-term strategic plans to facilitate incremental successes and transparency, inclusive conference planning, and leveraging disciplinary communities of practice to connect national networks and foster inclusive leadership. We hope these case studies help readers imagine and explore the possibilities for inclusive excellence in their disciplinary societies.

6

Bringing the IDEA to External Communications

DaJoie R. Croslan, Charles B. Fenster, James M. Verdier, Stephen A. Gallo, and Scott L. Collins

About AIBS

The American Institute of Biological Sciences (AIBS) is a society of societies with over 115 member organizations that serves the community by publishing impactful science in the peer-reviewed journal *BioScience* (currently published by Oxford University Press).[1] In addition to its publication activities, AIBS provides professional development opportunities for scientists and students, advocates for the life sciences community in policy settings, educates the public, and provides peer review and advisory and advisory support services for science funders and others. AIBS achieves these aims in collaboration with its member societies and organizations as well as like-minded, nonmember organizations, with the aim of promoting the thoughtful use of science to inform sound decision-making. Although the organization no longer has individual members, it serves over 100,000 scientists, students, and others in the life sciences community through its member societies and organizations.

AIBS Diversity Plan

In 2021, AIBS formally published its diversity plan (Croslan et al. 2021), which described its commitments to bolstering inclusion, diversity, equity, and acceptance (IDEA) in the life sciences community. Central to the AIBS Diversity Plan is the notion that practices related to DEI be embedded within the organization's core activities of assessment, training, communication, and advocacy. To achieve this end, the organization formed a diversity committee, composed of members of the AIBS staff and its board of directors. The newly established committee was charged with continually evaluating progress toward measurable goals, supported by periodic reporting to the AIBS Board of Directors. Because of its broad membership of scientific societies and other organizations (e.g., field stations), we hoped that AIBS could serve as a valued community forum for discussing IDEA-related issues, thereby helping to establish best practices for the scientific societies and other organizations that together represent the life sciences community. Herein, we focus on the society's external communications goals.

External Communications Objectives

AIBS has served for decades as a locus for discussions of IDEA-related topics (e.g., Uriarte et al. 2007), but our 2021 diversity plan offered an opportunity to approach the issue with greater deliberation and to embed related practices within our institutional activities and reporting. With regard to AIBS's communications core activity, the AIBS Diversity Committee outlined the following objectives:

> We will incorporate more editorials and podcasts addressing DEI topic areas. These will highlight perspectives, as well as novel research, and will be authored internally and externally. We will also ensure that we have a broader representation of underrepresented minorities as authors of op-eds of all topic areas, not just DEI issues.
>
> (Croslan et al. 2021, 778)

To form a baseline for understanding the demographic representation among *BioScience*'s authorship, the committee determined that a survey of *BioScience*'s recent authors would be valuable. We elected to survey authors of manuscripts accepted in the most recent three years, with the aim of capturing primarily recent authorship while still providing a sufficiently large sample size. The insights gained from this survey would then be used to measure subsequent performance against AIBS's diversity goals. Notably, the survey described here marks only the first steps in an effort that will be pursued over the long term.

To launch the survey—which addressed gender, sexuality, race and ethnicity, institution type, country, and years since the last degree was awarded as well as specific authorship characteristics (survey details are discussed later in this chapter)—we published an editorial describing the effort:

> *BioScience* is presently conducting a survey to help us better understand several aspects of the diversity (gender, ethnicity, career stage, employment, geographic

> distribution) of our authorship. That understanding will form a baseline from which to measure our progress toward a future in which a diversity of voices and perspectives, reflecting that of society at large, is presented within the pages of *BioScience*.
>
> (Collins et al. 2021a, 671)

At the time of the initial survey, we elected not to survey authors of rejected manuscripts. Although the resultant data could provide valuable evidence of bias (or the absence thereof) in our editorial processes, a high proportion of *BioScience* submissions are deemed inappropriate on topical grounds. It was our assessment that the inclusion of all submitting authors would generate an unwieldy sample that was poorly representative of the *BioScience* community.

Expanding Coverage

In addition to helping us learn more about—and ultimately grow—the diversity of *BioScience*'s authorship, the plan also articulated a linked goal of expanding the society's external communications specifically related to IDEA topics. This objective reflects AIBS's position as an umbrella society that provides a forum for conversations about best practices and other issues of broad import in the life sciences community. Under this remit, *BioScience* publishes articles in categories that extend beyond traditional research publications and instead, provides opportunities to address scientific practice itself. These categories include our pedagogy-focused Education category, Biology in History, Thinking of Biology, In Their Own Words (an oral history series), book reviews, special reports, and several commentary categories that permit the publication of longer opinion pieces. Along similar lines, AIBS has developed additional communication channels, including social media accounts, the BioScience Talks podcast (conversational interviews on life sciences topics), and the BioScience Bytes web series (short articles aimed at the general public), among others. It was our hope that this variety of outlets could foster ongoing discussion of IDEA-related issues.

Importance of Academic Publishing

The importance of publication in respected, high-impact journals is well acknowledged across academic fields and institutions (e.g., Niles et al. 2020), with a practitioner's publication record often playing a considerable role in the attainment of tenure and other professional outcomes. Regrettably, academic publishing also reflects wider societal biases, as seen in lower citation rates, slower publication times, and rarer representation on editorial boards among members of communities that have been historically excluded (Liu et al. 2023).

Given the role of publishing success on other measures of professional achievement, inequities in scholarly communication may constitute a covert barrier in other career areas for members of groups that have been historically excluded from science. In light of this potential harm, thoughtful consistent measurement and mitigation of biases are key to addressing broader inequities within the academy.

The value of a more diverse academy extends far beyond educational settings. Researchers who are members of historically excluded groups may be better positioned to addressing pressing societal issues that disproportionately affect underserved communities (Yitbarek et al. 2021). Air pollution and its harmful effects on human health provide a useful example. Air pollution contributes to higher rates or mortality among Black Americans than among members of other groups (Geldsetzer et al. 2024), and issue that warrants more intense study and remediation. In this sense, greater diversity among researchers has the potential to support broader social justice aims, with benefits accruing to society as a whole.

Author Demographic Data Collection: Past and Present

Although biases in journal publishing have been previously discussed in the literature, this work is most often undertaken with regard to editorial board membership, and often with a specific focus on geographical (Collins and Verdier 2018; Espin et al. 2017) or gender (Budden et al. 2007) inequities. This focus may reflect the challenges of collecting self-reported demographic data in comparison with, for instance, drawing on publicly available geographic information from participants' institutions. The failure to deliberately address other axes of potential bias, such as those related to race, creates a risk of overlooking the particular barriers faced by the most marginalized members of society (Crenshaw 1989).

Interest among publishers in examining a deeper set of diversity variables in academic publishing grew rapidly following the death of George Floyd and the consequent civil unrest, which was shortly followed by the launch of the

Royal Society of Chemistry's Joint Commitment for Action on Inclusion and Diversity in Publishing in June 2020. However, it was not until April 2022 that the Joint Commitment published a standardized set of questions for diversity data collection (RSC n.d.).

The Joint Commitment's three standardized questions (on gender, ethnic origins, and race) are now available within multiple manuscript submission management software packages (including ScholarOne, used by *BioScience*). This allows the anonymous collection of data from reviewers, authors, and editors. The streamlined collection process and ease of implementation will allow for the creation of the large data sets needed to interrogate the demographic characteristics of contributors to thousands of journals. However, at the time of this case study's initiation, no such schema was in wide use. Rather than await the development of a standardized approach, we instead elected to develop our own. It is our belief that increasing IDEA in the life sciences is crucial for improving opportunities for individual scientists from historically excluded groups but also for the benefit of society at large. It is our belief, outlined in AIBS's diversity statement, that science is strengthened by the open exchange of diverse perspectives and ideas, which lent urgency to our efforts. It was our hope that this baseline of historical data could provide a basis for the measurement of our future IDEA-expansion efforts.

Tracking IDEA Publications

Although our own efforts in this respect are modest, we are unaware of other attempts to track the quantity of IDEA-related content published through scientific society external communication channels. By building this effort into our reporting structure to the AIBS Diversity Committee and broader Board of Directors, it was our hope that a structure of continuous review of our performance would establish IDEA coverage as a perpetual area of focus, not subject to the drift that often affects editorial priorities over the passage of time and with changing personnel.

A Survey of Authors

The case study described here differs from others described in this book because its principal survey outlines the collection of data not from members of a scientific society but rather, from the contributing authors to its journal,

BioScience. The journal publishes primarily review and synthesis articles of broad ecological and conservation relevance, with primary research included only in categories relating to biology education and citizen science. A particular challenge for our author demographic survey lay in the transitory nature of relationships between journals and authors. Although *BioScience* has benefited from authors who repeatedly engage with the journal (Gallo et al. 2022a), many authors interact with the society and its publications only infrequently. In addition, authors who contributed to articles with multiple coauthors may have had limited (or no) direct interactions with AIBS and journal staff. To address this, we published an editorial (Collins et al. 2021b) announcing the survey's launch, with the aim of building community awareness of and interest in our efforts.

The secondary track of our case study is focused on increasing our quantity of IDEA-related publications. Crucial to AIBS's society-of-societies role is the inclusion of publication—within and outside of *BioScience*—of topics with a broad interest to the biological sciences community, particularly those describing the field itself. Given this positioning, AIBS was, and is, well placed to provide coverage of and commentary on topics (such as IDEA) that are important to our member societies and organizations as well as to others in the life sciences community.

Contributions

During the period described in our case study, *BioScience* was led by two editors-in-chief who appear as authors of this chapter. Scott L. Collins began his tenure as *BioScience*'s editor-in-chief in 2016. After he stepped aside at the end of 2022, Charles B. Fenster took over the helm of the journal. Dr. Fenster had previously held numerous positions on the AIBS Board of Directors, including as its president. A lengthy transitional period between editors, as well as continuity in AIBS staff (James M. Verdier served as senior editor for the duration of the case study), permitted the conduct of and reporting on our survey without interruption. DaJoie R. Croslan serves as chief operations and diversity officer at AIBS. At the time this work was conducted, Stephen A. Gallo was the chief scientist of AIBS. The small size of the AIBS organization and overlapping work objectives facilitated close collaboration in the efforts described herein.

Author Survey Approach

Because authors interact infrequently with the journal, we selected three years of authorship as a baseline period. Our hope was that by surveying a broad swath of authors even a low response rate would allow for statistically meaningful analyses of the data. In our publication of the results, we described our general survey approach, which was sent to all authors represented within the survey period.

> We compiled a list of 2,028 email addresses of potential respondents from our ScholarOne peer-review system. The list of potential respondents included authors of *BioScience* articles accepted between 1 January 2018 and 31 December 2020. Because some authors had multiple email addresses and because it was not feasible to limit one email address per individual, the respondents were given instructions in an invitation email to fill out the form only once. Our survey was sent out 11 August 2021, with 96% of the emails successfully delivered. One reminder email was sent out to the list two weeks later, and the survey was closed in early October 2021.
>
> (Gallo et al. 2022a, 1044)

Survey Details

Because of the factors described previously as well as our interest in comparing the ways in which authorship characteristics (role, article type, number of articles) may interact with demographics, we asked a series of questions beyond standard demographic queries. In general, our question set was developed to conform with previous surveys conducted by the NSF (2019) and prior internal surveys conducted by AIBS. Google Forms was selected as our data collection platform to align with AIBS's prior practices and its likely familiarity among potential respondents.

> We developed a survey using Google Forms to better understand the baseline demographics of *BioScience* authors, asking standard identity questions about

> gender, sexuality, race and ethnicity, institution type, country, and years since the last degree was awarded [as a rough estimate of career stage]. We also collected information about the respondents' activities with *BioScience*—namely, their most recent authorship role (first, last, middle), how many of their articles had been accepted by *BioScience* in the last three years, and the type of article (e.g., Overview, Education, Viewpoint). We provided predetermined multiple-choice answers for all questions (for race or ethnicity and article type, the respondents could mark all that apply). For demographic questions, the respondents were also given a "prefer not to answer" option and for sexuality, race, and institution; an "other" option was provided with an open text field to submit an answer not represented in the choices. A total of nine questions were developed, including an optional open text field at the end for any general comments. We estimated that the survey would take five minutes to complete. In the survey's introduction, we outlined the purpose of the survey, the anonymity and privacy of the data collection, the potential benefits in participating, and a consent to the analysis of the anonymized data collected. The respondents were also given an email to contact if they had any questions.
>
> (Gallo et al. 2022a, 1044)

To protect respondents from potential bias, our initial authorship survey was developed with the aim of ensuring that participation was entirely voluntary and anonymous, as described and communicated to all potential respondents. In addition, to help us calibrate our ongoing efforts to ensure greater participant comfort, we included the free-text box described above in which respondents could provide anonymous feedback on the survey itself as well as on the broader data collection effort. The free-text box entries were not formally analyzed, given their general brevity and broad topical range.

A potential limitation of our broader approach relates to our use of "years since last degree" as a proxy for age and career stage. In developing our survey, we sought to limit the total number of questions, to both limit the burden on participants' time and increase the participation rate. Our approach may have failed to adequately reflect nontraditional academic careers, which could have been elucidated through a comparison of age and career stage. In addition, the inclusion of questions concerning career focus (e.g., research or teaching focused) may have provided an opportunity for additional analysis and insight.

Response Rate

Also in our publication of the survey results, we described our response rate, which was broadly in line with rates from similar surveys previously conducted by AIBS.

> A total of 283 responses were recorded from the survey (13.9% response rate), which compares favorably with similar surveys conducted by AIBS. Although all of those respondents completed the survey, only 35 made a general comment in the optional text box, with their responses ranging from those about how the survey was implemented to comments on the *BioScience* journal itself.
>
> (Gallo et al. 2022a, 1045)

In general, although the response rate was deemed satisfactory, it is worth considering that there may be sampling biases attributable to career stage, author values and interest in IDEA issues, and other factors. We attempted to address some of these concerns by including questions related to career stage and authorship role, but as is typically the case with surveys of this type, it is all but certain that our sample is imperfectly representative of our authorship.

Data Collection on IDEA Topical Coverage

In addition to our survey, we initiated tracking our coverage of IDEA-related issues across a variety of article types found in *BioScience,* including peer-reviewed articles as well as content published by the society outside of the journal, such as podcasts, press releases, and web items. The tracking process was conducted informally, with continuous input from the AIBS Diversity Committee on appropriate inclusion criteria. We began collecting data in January 2021 and maintained a list of publications with an IDEA-related focus. Because our recordkeeping spanned widely varied publication types (for instance, audio-only podcasts) traditional bibliometric approaches to classification were largely precluded. Instead, we followed the diversity committee's advice and included in our list any article deemed by staff to have substantial coverage of IDEA-related topics. For instance, an oral history published in *BioScience* in which the interviewee substantively discussed the challenges of entering the academy as a member of a historically excluded group would be included, but an editorial that made only passing mention of diversity would not. It is our expectation that as IDEA values become further enmeshed in the life sciences these topics will be more frequently discussed in articles with multiple foci beyond IDEA considerations. Our hope was to capture this phenomenon in our data collection approach.

Summary Findings of Racial and Gender Demographics

We have published elsewhere in greater detail (Gallo et al. 2022a), but our basic findings for racial and gender demographics were broadly in line with our expectations that *BioScience* would exhibit demographic characteristics similar to the broader life sciences community. This result is, however, far short of our long-term aspiration that *BioScience*'s authorship will more closely reflect that of the general public, both in the United States and internationally. It should be noted that our sample reflects an international author population as roughly two-thirds of our respondents reported working in the United States or Canada at the time of survey completion. Demographic comparisons of our respondents with the US population should be considered with this in mind. In part because of the very low numbers of respondents from some countries, a country-level demographic analysis would have risked de-anonymizing respondents and was therefore infeasible. The following results are depicted in Figure 6.1, a reprinting of a figure from our special report.

> Black scientists were 3.5% of the PhD recipients in the biological sciences, despite representing 13% of the US population, according to 2018 National Science Foundation (NSF) data (NSF 2019). Although our finding that 4% of *BioScience* authors identify as African American or Black seems in line with the proportion of Black biologists in the United States, this figure is woefully lower than the general population. A similar case exists for Latinx scientists, who make up 6% of biology PhD recipients and 10% of our sample but roughly 19% of the US population. More work is needed throughout the pipeline, and all scientific entities have a part to play. Moreover, the same NSF (2019) study indicated that 53% of biology PhDs were granted to female scientists, and women make up 50.1% of the general population. Although 46% of the surveyed *BioScience* authors identified as female, this percentage is slightly lower than both of these other statistics, suggesting more work still needs to be done to achieve [gender] proportionality.
>
> (Gallo et al. 2022a, 1048)

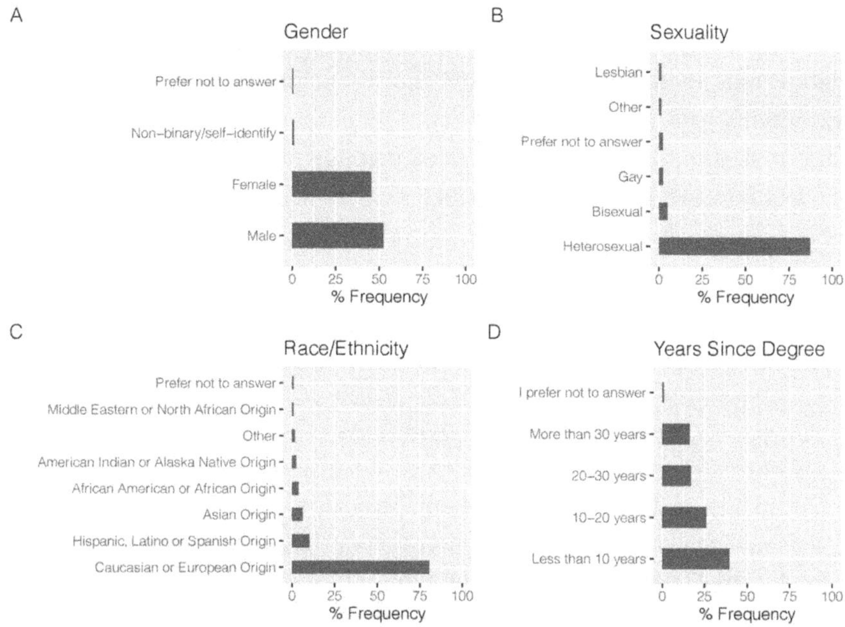

Figure 6.1 Author demographics. (a) gender, (b) sexuality, (c) race or ethnicity, and (d) years since degree. Reprinted from Gallo et al. 2022a.

These results will be particularly interesting to compare with those of peer journals as they publish the results of their ongoing demographic data collection. It must be noted that results will likely vary in accordance with subdiscipline and journal focus. For instance, according to NSF (2019) data, only 3.2 percent of graduate students in natural resources and conservation identified as Black, versus 5.6 percent in the biological and biomedical sciences. Future cross-journal comparisons will therefore require close consideration of journals' specific disciplinary focus.

Interaction between Authorship Characteristics and Demographics

We were interested in elucidating any possible interactions between author demographics and characteristics (i.e., position in the author list, number of articles published, and article type), in part because these characteristics may relate to publication prestige (e.g., position in the author list) or the perceived significance of a publication (e.g., article type). To address this, we developed models to describe interactions between gender, sexuality, and race or ethnicity with authorship characteristics, but we found no statistically discernible relationships in this regard. We did, however, find statistically discernible interactions with respect to author age. To avoid multicollinearity resulting from correlations between age and other demographic variables, we separated our data into two groups: one in which authors had received a degree within the last twenty years (n = 186) and one in which more than twenty years had passed since the author's last degree award (n = 94).

> Using the full data set, we conducted separate regression models to explore only the effect of age on all three authorship characteristics. The results are listed in supplemental tables S7–S9. All three models were statistically discernible, although the amount of variance they explained was small (3–5%). Nevertheless, age was a discernible predictor of all three authorship characteristics. We then plotted the distributions of the number of years since their degree alongside first and middle authorship compared with last authorship, which suggested an older age for last authors []. Examining the number of years since their degree as an interaction with article type suggests that the authors that published article types other than Overviews or Forums are slightly older []. A plot of the distribution of the number of years since their degree as a function of the number of articles suggests that the authors of multiple articles are slightly older than those of only one article []. It should be noted that the authorship characteristics data for all three plots were somewhat skewed (most of the authors were not last authors). Nevertheless, it seems likely that older authors in our sample were more likely to have published more than one article in *BioScience* over this period, were more likely to be a last author of an article, and were more likely to publish article types that were not Overviews or Forums.
>
> (Gallo et al. 2022a, 1046–48; contains the figure and supplements cited)

It is our view that these results largely reflect publishing conventions within the life sciences. Specifically, senior scientists are more likely to occupy the position of last author, as was reflected in our results. In terms of article type interactions with career stage, it seems reasonable to infer that article types focusing on the practice of biology itself (e.g., Thinking of Biology, Biology in History) may be more popular among senior or emeritus academics with less active ongoing research. It is also notable that some results may have been skewed by AIBS-affiliated authors with numerous publications during the period (e.g., editors and staff).

Comparability with Other Journals and Publishers

Despite the importance of comparison with peer journals, as we undertook our initial authorship survey we found very few sources of data. We highlighted this challenge in a *BioScience* special report describing our results:

> There are surprisingly few publicly available data sets, and the results are not standardized (Else and Perkel 2022) and have often been focused on relatively few demographic characteristics. For AAAS *Science* journals' authors and reviewers, those identifying as Black represented less than 1% of the total, although this number is hard to assess because of missing data (AAAS 2020). For cardiology journals, a recent tally suggested that only 2% of their authors are Black, although these results were not self-identified (Else and Perkel 2022). For American Chemical Society journals, Black corresponding authors represented 1% of the total (ACS 2021). In our results, Black authors represent 4% of the journal's total. Similarly, the reported levels of Latinx authors in these journals appear significantly lower than the 10% that are represented in our sample. In terms of gender, the male authors and reviewers for AAAS Science journals were found to outnumber their female authors 2.3:1, whereas in American Chemical Society journals, the corresponding published female authors represented 23% of the total. In our sample, the male authors of *BioScience* barely outnumbered the female authors (1.2:1).
>
> (Gallo et al. 2022a, 1048)

These results may provide an early indication that *BioScience* is performing well in some diversity metrics, at least in comparison with journals that published data reflecting authorship over a similar time period. However, differences between discipline-specific demographic characteristics and the lack of standardized data with which to compare our results likely preclude any meaningful comparative inferences at this time. It is our hope that standardized demographic data collection methods and schema, following the Joint Commitment's recommendations, will allow more useful comparisons of *BioScience*'s authorship diversity with that of its peer journals. We are compelled to note that even in the absence of detailed comparisons, our results and others' demonstrate a clear and urgent need to train and retain greater numbers of scientists from underrepresented groups.

IDEA Publication Tracking Results

As is described above, our tracking began in January 2021 and continues until the present (mid-July 2024, as of this writing), reflecting a period of approximately three years, seven months. During this period, AIBS published a total of ninety-seven IDEA-related pieces, including press releases, web-only articles (BioScience Bytes), podcasts, commissioned features, peer-reviewed articles, commentaries, special reports, and editorials.

Many of these publication channels, such as BioScience Bytes (representing 10 publications) were redirected toward our IDEA-related efforts in 2021 and constitute outreach that would not have occurred in the absence of our stepped-up IDEA program. For these types of publications, comparison with a similar prior period is challenging. However, for publication types that existed largely in their present form before the publication of our diversity plan, we saw broad improvement in our IDEA coverage. For instance, in the period of three years, seven months preceding January 2021, we published only five IDEA-related podcasts, in comparison with fifteen published in the study period. Similarly, during the period prior to the case study, we published only three press releases on IDEA topics compared with six published during the case study period, which represents a doubling of coverage.

We showed more modest gains in the peer-reviewed literature, with thirty peer-reviewed articles on topics related to IDEA during the study period and about half of those clustered around education-related topics. In the previous period, spanning June 2017 through January 2021, we published twenty-seven IDEA-related peer-reviewed articles. Although the growth in topical coverage is modest, recent articles tended to spend more space addressing IDEA topics. It is notable that peer-reviewed articles are typically unsolicited, and our results here demonstrate that *BioScience* has long been, and continues to be, a valued outlet for vetted publications on IDEA topics. A broader analysis that focuses on multiple society journals undertaking similar efforts could yield interesting results and help discern the degree to which topical trends are attributed to shifting values within the field or the IDEA-related efforts of specific publishers.

Discussion and Reflection

Our author survey offers a possible rubric for academic publications interested in analyzing the baseline demographic characteristics of their authorship. Although the integration of demographic queries within platforms such as ScholarOne and Editorial Manager will provide societies and other publishers with a sound source of data on their present and future authorship, a survey such as ours is useful in establishing an understanding of an authorship's baseline characteristics *prior to* the initiation of, for example, an outreach program aimed at increasing demographic diversity among authors.

It is our hope that our early effort in journal demographic data collection can be valuable to our member societies and the broader scientific community in their continued efforts to increase IDEA within the sciences. We hope that this work will also serve to the benefit of scientists from groups that have been historically excluded and to society writ large.

Further Questions, Future Directions

Numerous signatories of the Joint Commitment for Action on Inclusion and Diversity in Publishing and many others have adopted the program's standardized set of questions for collecting demographic data (RSC n.d.). As described above, *BioScience* continues to collect data following this question set.

The standardized approach is valuable and was developed thoughtfully. In developing their questions, the Joint Commitment conducted a survey with 1,174 respondents with the aim of ascertaining their "perceptions of the representativeness of the options provided, and their comfort with sharing race and ethnicity demographic information when in the role of editorial board member vs. reviewer vs. author" (RSC n.d.).

While we appreciate the robustness of this stakeholder engagement and the focus on respondents' comfort in answering the questions, it remains to be seen whether the question set developed—which addresses gender, ethnic origins and ancestry, and racial self-identification—will provide a sufficiently broad spectrum of data across various axes of diversity and possible bias. For instance, a respondent to our author survey noted that disability was an important area of diversity missed in our initial survey (Gallo et al. 2022a).

Further, it remains to be seen whether narrowly focused schema will be sufficient in addressing the crucial role of intersectionality. Approaches to alleviating bias that fail to consider multiple, interacting identities (e.g., gender, sexuality, class, and race) will tend to miss the unique forms of subordination faced by those who are members of multiple historically excluded groups (Crenshaw 1989). If data collection is confined to a limited set of characteristics, there is a risk that attempts to prevent marginalization will fail to address the systemic barriers faced by many individuals. This may require particular attention in the field of academic publishing, a sphere in which biases and covert censorship are often veiled within systems of peer review (Strauss et al. 2023).

We believe that such concerns warrant attention, and we are presently considering the possibility of periodic supplemental surveys with the aim of broadening our understanding of our community's diversity. It will be particularly important to ensure that disability-related survey schemas are well aligned with the appropriateness and comfort objectives described in the Joint Commitment's gender, ethnicity, and race schema. Although standardized schemas exist for disability data collection (e.g., CDC 2024), such question sets may be inappropriate in the context of academic publishing. Members of the Joint Commitment have suggested that additional formally standardized schema are on the way ("Additional diversity dimensions will be considered as the program proceeds" [Hill-Parks 2023]), but there is no present timeline for their circulation. AIBS and similar scientific societies will need to carefully weigh the risks and benefits of awaiting the establishment of an industry-wide schema against the need for urgent action to improve IDEA within the sciences.

Applications beyond AIBS: Expansion of IDEA Coverage

One of the key findings of our effort was the value of nontraditional communication channels in addressing IDEA-related issues. Diversity-related communication has frequently taken the form of commentaries and personal stories, with social media playing a crucial role (Hinton Jr. and Lambert 2022). We found that nontraditional publications offer societies and other actors a valuable opportunity to quickly shift community focus to crucial issues. This is an important avenue of communication that should not be overlooked while pursuing Hinton Jr. and Lambert's (2022) call for the broader inclusion of quantitative and solution-focused perspectives in the IDEA literature. Alongside this note of encouragement to our peers, we recognize that many scientific societies have minimal to no staff and limited financial resources at their disposal, and undertaking efforts such as ours may prove difficult for many organizations. Nonetheless, as described by Shiffman et al. (2022), efforts to amplify the voices that have been historically excluded from science through nontraditional channels represent a relatively easy to implement approach to improving IDEA.

To this end, scientific societies wishing to broaden their coverage are urged to consider repurposing nontraditional publication channels. For instance,

AIBS used its BioScience Bytes series of brief, online-only articles to highlight numerous heritage months throughout 2023 (e.g., Black History Month). These articles also provided an opportunity for engagement with representatives from the societies that constitute AIBS's membership, such as in a piece celebrating Hispanic Heritage Month (Brito et al. 2023).

In addition to repurposing or developing nontraditional communication formats, societies may also consider increasing the accessibility of publication in their journals. To this end, AIBS negotiated with *BioScience*'s publishing partner, Oxford University Press, to remove publishing fees from a number of article types that have often been used in the discussion of IDEA topics. We encourage other societies to frequently revisit their publications' financial models and, where possible, consider whether minor adjustments to a model may produce outsize improvements in areas of coverage without negatively affecting the publication's financial sustainability.

Future Work at AIBS

We will continue our tracking and reporting of IDEA-related topical coverage, with an eye to maintaining its relevance to the community. We will also continue to explore the possibility of expanding our available communication channels in hopes of developing formats that provide outlets that serve both author and audience needs.

To increase the diversity of *BioScience*'s authorship, our primary effort will lie in the diversification of our editorial board. Research has demonstrated that diversity among editorial board membership results in greater diversity among authors (Goyanes and Demeter 2020). To this end, we have initiated a demographic survey of our approximately 100 editorial board members, and we are working to develop outreach and candidate identification approaches to ensure that our editorial board recruitment is well aligned with AIBS's values and priorities. One approach may lie in the adaptation of the cluster hires concept promoted by Madzima and MacIntosh (2021). By recruiting several members of historically excluded groups simultaneously, there is the potential to foster a more inclusionary and supportive peer environment. It is crucial that these efforts additionally consider the role of intersecting identities and the unique barriers to those facing subordination across multiple interacting axes.

In addition to our forthcoming editorial board efforts, we look forward to comparisons of our now continuously collected author demographic data with that collected in our initial demographic survey. Although the question schemas used are not identical, sufficient similarities exist to permit future analyses. More expansive data sets will also provide the opportunity to partition the data to allow comparisons among authors of accepted and rejected manuscripts that were deemed topically appropriate. We will continue to report the results of our efforts within AIBS's internal reporting structure and to the broader life sciences community.

Conclusions

Despite early successes in improving the topical mix in our communication channels, significant challenges remain, particularly with regard to broadening representation in *BioScience*'s authorship. To some extent, a lack of racial and ethnic diversity may be a function of *BioScience*'s traditional focus on reviews and syntheses rather than primary research publications, which may skew our authorship toward a later-career cohort. As noted by Gallo et al. (2022a), the journal's younger demographic is relatively more diverse, so "focused outreach to early-career researchers will be key in ensuring that *BioScience* is home to the diverse perspectives that will shape science's future."

We hope that our success in increasing the society's external communications relating to IDEA topics will be of use to the broader community. It is particularly notable that this effort relied almost entirely on resources that were extant before the initiation of the AIBS Diversity Plan, and came with the strong support of our publisher, Oxford University Press. The formats and channels we used (and continue to use) to share IDEA-related information were mostly already in place at the organization and with our publisher, ready for refocusing and repurposing to help reach new objectives.

We highlight the importance of deeply embedding programs such as ours within the operating model, overarching philosophy, and reporting structure of the broader organization. In our case, setting out clear, concrete objectives in our diversity plan—together with prearranged periodic reporting to our diversity committee and board of directors—ensured that our focus did not falter. We encourage scientific societies and other organizations to build continual assessment into their own IDEA-related programs—this work is crucial to our disciplines and to society as a whole. It deserves our unwavering attention and effort.

Note

1 We are indebted to the members of the AIBS Diversity Committee and Board of Directors as well as AIBS's senior leadership for their steadfast commitment to these efforts. We are likewise grateful to the LED-BIO team for the opportunity to share our work with the community.

References

AAAS (American Association for the Advancement of Science). 2020. "Baseline Assessment of Demographic Representation in AAAS/*Science* Functions." AAAS, October 22. www.aaas.org/sites/default/files/2020-10/Baseline%20Assessment%20of%20Demographic%20Representation%202020%20Public_20201020_508_updatedpublicslide.pdf.

ACS (American Chemical Society). 2021. "Announcing the ACS Publications Diversity Data Report 2021." *ACS Publications*, December 21. https://axial.acs.org/publishing/announcing-the-acs-publications-diversity-data-report-2021.

Brito, S., C. Miranda-Aldaco, and V. A. Segarra. 2023. "Celebrating Hispanic Heritage Month at AIBS & Goucher College." *American Institute of Biological Sciences*, September 15. https://www.aibs.org/news/2023/230915-hispanic-heritage-month.html.

Budden, A. E., T. Tregenza, L. W. Aarssen, J. Koricheva, R. Leimu, and C. J. Lortie. 2007. "Double-Blind Review Favours Increased Representation of Female Authors." *Trends in Ecology and Evolution* 23 (1): 4–6. https://doi.org/10.1016/j.tree.2007.07.008.

CDC (Centers for Disease Control and Prevention). 2024. "Disability and Health Promotion: Disability Datasets." Last modified July 15. https://www.cdc.gov/ncbddd/disabilityandhealth/datasets.html.

Collins, S. L., and J. M. Verdier. 2018. "Editorial Boards Must Be Internationally Representative." *BioScience* 68 (4): 235. https://doi.org/10.1093/biosci/biy030.

Collins, S. L., D. R. Croslan, and S. A. Gallo. 2021a. "Diversity Survey Results." *BioScience* 71 (11): 1115. https://doi.org/10.1093/biosci/biab118.

Collins, S. L., S. A. Gallo, and J. M. Verdier. 2021b. "The *BioScience* Author Survey." *BioScience* 71 (7): 671. https://doi.org/10.1093/biosci/biab069.

Crenshaw, K. 1989. "Demarginalizing the Intersection of Race and Sex: A Black Feminist Critique of Antidiscrimination Doctrine, Feminist Theory and Antiracist Politics." *University of Chicago Legal Forum* 1989 (1): 139–67. https://chicagounbound.uchicago.edu/uclf/vol1989/iss1/8.

Croslan, D. R., G. Chavarria, R. Nakamura, J. L. Pandey, S. T. A. Pickett, C. Poodry, K. Skogen, M. Zaka, and S. A. Gallo. 2021. "AIBS Diversity Plan." *BioScience* 71 (8): 787–9. https://doi.org/10.1093/biosci/biab026.

Else, H., and J. M. Perkel. 2022. "The Giant Plan to Track Diversity in Research Journals." *Nature* 602: 566–70. https://doi.org/10.1038/d41586-022-00426-7.

Espin, J., S. Palmas, F. Carrasco-Rueda, K. Riemer, P. E. Allen, N. Berkebile, K. A. Hecht, K. Kastner-Wilcox, M. M. Núñez-Regueiro, C. Prince, C. Rios, E. Ross, B. Sangha, T. Tyler, J. Ungvari-Martin, M. Villegas, T. T. Cataldo, and E. M. Bruna. 2017. "A Persistent Lack of International Representation on Editorial Boards in Environmental Biology." *PLOS Biology* 15 (12): e2002760. https://doi.org/10.1371/journal.pbio.2002760.

Gallo, S. A., J. M. Verdier, and S. L. Collins. 2022a. "Representation in *BioScience* Authorship." *BioScience* 72 (11): 1044–9. https://doi.org/10.1093/biosci/biac056.

Geldsetzer, P., D. Fridljand, M. V. Kiang, E. Bendavid, S. Heft-Neal, M. Burke, A. H. Thieme, and T. Benmarhnia. 2024. "Disparities in Air Pollution Attributable Mortality in the US Population by Race/Ethnicity and Sociodemographic Factors." *Nature Medicine* 30: 2821–9. https://doi.org/10.1038/s41591-024-03117-0.

Goyanes, M., and M Demeter. 2020. "How the Geographic Diversity of Editorial Boards Affects What Is Published in JCR-Ranked Communication Journals." *Journalism and Mass Communication Quarterly* 97 (4): 1123–48. https://doi.org/10.1177/1077699020904169.

Hill-Parks, E. 2023. "Paving the Way to Increase Diversity in Journals and Research." *Elsevier Connect*, March 23. https://www.elsevier.com/connect/paving-the-way-to-increase-diversity-in-journals-and-research.

Hinton Jr., A., and W. M. Lambert. 2022. "Moving Diversity, Equity, and Inclusion from Opinion to Evidence." *Cell Reports Medicine* 3 (4): 100619. https://doi.org/10.1016/j.xcrm.2022.100619.

Liu, F., T. Rahwan, and B. AlShebli. 2023. "Non-White Scientists Appear on Fewer Editorial Boards, Spend More Time under Review and Receive Fewer Citations." *Proceedings of the National Academies of Sciences* 120 (13): e2215324120. https://doi.org/10.1073/pnas.2215324120.

Madzima, T. F., and G. C. MacIntosh. 2021. "Equity, Diversity, and Inclusion Efforts in Professional Societies: Intention versus Reaction." *The Plant Cell* 33 (10): 3189–93. https://doi.org/10.1093/plcell/koab186.

Niles, M. T., L. A. Schimanski, E. C. McKiernan, and J. P. Alperin. 2020. "Why We Publish Where We Do: Faculty Publishing Values and Their Relationships to Review, Promotion, and Tenure Expectations." *PLOS ONE* 15 (3): e0228914. https://doi.org/10.1371/journal.pone.0228914.

NSF (National Science Foundation). 2019. *Women, Minorities, and Persons with Disabilities in Science and Engineering: 2019.* Special Report NSF 19-304. National Science Foundation. https://ncses.nsf.gov/pubs/nsf21321/data-tables.

RSC (Royal Society of Chemistry). n.d. "Diversity Data Collection in Scholarly Publishing." Accessed 2024. https://www.rsc.org/policy-evidence-campaigns/inclusion-diversity/joint-commitment-for-action-inclusion-and-diversity-in-publishing/diversity-data-collection-in-scholarly-publishing.

Shiffman, D. S., T. A. Álvarez, C. W. Bangley, et al. 2022. "What Can Professional Scientific Societies Do to Improve Diversity, Equity, and Inclusion: A Case Study of the American Elasmobranch Society." *Frontiers in Education* 7: 842618. https://dx.doi.org/10.3389/feduc.2022.842618.

Strauss, D., S. Gran-Ruaz, S. Gran-Ruaz, M. Osman, M. T. Williams, and Sonya C. Faber. 2023. "Racism and Censorship in the Editorial and Peer Review Process." *Frontiers in Psychology* 14. https://dx.doi.org/10.3389/fpsyg.2023.1120938.

Uriarte, M. S., H. A. Ewing, V. T. Eviner, and K. C. Weathers. 2007. "Constructing a Broader and More Inclusive Value System in Science." *BioScience* 57 (1): 71–8. https://doi.org/10.1641/B570111.

Yitbarek, S., K. Bailey, S. Tyler, J. Strickland, M. McCary, and N. C. Harris. 2021. "Inclusive Sustainability Approaches in Common-Pool Resources from the Perspective of Blackologists." *BioScience* 71 (7): 741–9. https://doi.org/10.1093/biosci/biab052.

7

Expanding the National Research Ecosystem through Research Development: A Case Study of the National Organization of Research Development Professionals Consultants Program

Kimberly Eck

The National Organization of Research Development Professionals (NORDP), a professional organization for those working to grow resources for research, launched the NORDP Consultants Program in 2021.[1] The program is dedicated to expanding our national research ecosystem by providing research development services to MSIs and emerging research institutions (ERIs) at no cost to the institution. Launching the NORDP Consultants Program was a significant undertaking for NORDP as an all-volunteer organization. This case study, authored by the inaugural senior director of the program, documents the genesis of and motivation for creating the program, explores program implementation considerations for launching a national initiative, and shares the resources needed, tools developed, challenges encountered, and mitigation strategies used. This is an active, ongoing initiative; some engagements are completed, others are underway, and more are planned. As such, this case study summarizes the outcomes of early engagements and feedback from MSI and ERI participants that have been completed as of fall 2024.

Engaging MSIs and ERIs in the national research ecosystem is essential to addressing the inequitable distribution of federal research funding, which perpetuates regional imbalances and limits the impact of efforts to broaden participation in the STEM workforce (Quider and Blazey 2023). Increasing research funding at MSIs and ERIs strengthens science through the variety of ideas (Hofstra et al. 2020). Further, research funding can support paid research opportunities for undergraduates, which can increase student retention rates in

STEM, help students develop a sense of scientific identity, and provide students with needed financial support (Quider and Blazey 2023).

In alignment with NORDP's mission to increase the capacity for and impact of research, scholarship, and creative endeavors by promoting excellence in the practice and profession of research development, the NORDP Consultants Program aims to have a national impact. This case study explores the role of a professional society in creating a more national inclusive STEM research community by increasing the MSIs and ERIs capacity to participate. The key research questions to be answered throughout this case study include:

1. What were the motivations and national context for NORDP to create the NORDP Consultants Program?
2. What challenges and key decisions were navigated during the initial launch of the NORDP Consultants Program?
3. What resources, tools, and mitigation strategies were developed and implemented to ensure the success of the NORDP Consultants Program?
4. What are the outcomes of the program to date?
5. What are the lessons learned and insights from this experience that may be generalizable to other professional or scientific organizations seeking to implement similar programs?

Importantly, this case study is being conducted while the program is active and continues to evolve. These insights are a snapshot of the program as it stands in summer 2024.

Background

About Research Development and the National Organization of Research Development Professionals

Research development is a function that seeks to grow research or increase the research reputation of an institution (Eck and Roney 2022). Regardless of title or position, most research development professionals engage in finding funding opportunities, facilitating collaborations, providing proposal development, and capturing and monitoring proposal and award metrics (Eck and Roney 2022).

Established in 2010 in response to increased research development investment by institutions of higher education (IHEs), NORDP coordinates and promotes the professional development of research development personnel. NORDP's mission is to increase the capacity for and impact of research, scholarship, and creative endeavors by promoting excellence in the practice and profession of research development. As of summer 2024, NORDP has more than 1,400 members serving in research development roles primarily in IHEs. NORDP is an all-volunteer organization with more than 100 highly engaged volunteers whose effort and expertise are deployed across diverse programming and services through a robust and appropriately complex organization infrastructure. NORDP is led by a fifteen-person board of directors and maintains eight committees populated by volunteers. Each committee is led by co-chairs who serve staggered terms to ensure continuity in the organization's essential functions. The NORDP board coordinates closely with these volunteers to advance critical projects and innovative programming, including the NORDP Consultants Program.

National Trends in Research and Research Development

There are large inequalities in the allocation of research funding, and this disparity is growing. As described by Pai et al. (2024), "researchers in IHEs must compete for grants to conduct research, including to support their salary, students, and research supplies. These are cyclical cumulative effects—more grant funding leads to more research and publications, which attracts more talent, more opportunities, and more grants." The National Science Foundation (NSF) conducts an annual census of US IHEs that spend at least $150,000 on research and development in a fiscal year. IHEs that report to the Higher Education Research and Development (HERD) Survey range from research-intensive institutions (considered research 1 [R1] doctoral universities with very high research activity in the Carnegie Classification system) and large academic medical centers to MSIs such as historically Black colleges and universities (HBCUs). The 150 IHEs with the highest total research expenditures reported to the HERD Survey in 2022 are primarily R1 and medical centers that have experienced a 70 percent increase in research funding since 2010. The over 500 remaining IHEs have experienced a 5 percent increase in total research expenditures over the same time period and are by definition ERIs (Pai et al. 2024).

From 2012 to 2015, federal funding (the largest source of research expenditure funding) experienced a temporary dip during the recession years following the American Recovery and Reinvestment Act (Pub. L. No. 111-5, 123 Stat. 115 [2009]) funding. In response, the top 150 IHEs used institutional funds, which grew from $9.9 billion in 2010 to $21.4 billion in 2022, to increase their total research expenditures. The remaining IHEs also experienced growth in institutional funds but at a much slower pace from $2 billion in 2010 (down sharply from $2.9 billion in 2009) to $3.1 billion in 2022. It was not until 2021 that these IHEs recovered to their pre-recession high of institutional funds (NCSES 2022).

There are many complex factors that interact cyclically with funding trends. MSIs have reported that limitations of their research-support infrastructure and inequitable partnerships are two key areas that contribute to this disparity (Pai et al. 2024). MSIs and ERIs have fewer sources of institutional funds on which to draw and relatedly fewer research development professionals. For example, while the indirect cost recovery for the top 150 IHEs has increased from $13.2 billion to $20.6 billion from 2010 to 2022, it decreased for all other IHEs from $1.9 billion to $1.7 billion (NCSES 2022). From an analysis of research development job descriptions and a salary survey, most research development professionals are employed by R1 IHEs and academic medical centers (Eck et al. 2020; Preuss et al. 2019, 2020), which leaves MSIs and ERIs at a disadvantage when competing for research funding.

In summary, since 2010, the disparity in research funding has allowed the top 150 IHEs to buoy their research portfolios with institutional funds during years of lean federal funding and invest in their research infrastructure, including employing research development professionals. Thus, a self-replicating and increasing inequality in research funding has been created between the top 150 institutions and other IHEs, including all HBCUs, ERIs, and many MSIs.

State of the Literature on Increasing Research Capacity in MSIs and ERIs

The impact of research development within IHEs on advancing science and other scholarly activities, especially in MSIs and ERIs, is understudied and largely unpublished; there have been no large-scale evaluations in the United States. Eck and Roney (2022) note several challenges to studying research development, including "research development" not being a universally adopted term, a plethora of terminology and titles used to describe research development offices and positions, ongoing research development activities that existed before the term *research development* was coined that may have been framed as faculty development or similar, variations in scope of responsibilities from institution to institution and individual to individual, and an evolving practice of research development. Currently, there is no peer-reviewed journal dedicated to publishing the impact of research development. Although, one is under development, it has not released its first issue. What limited literature does exist is focused on presenting a description or evaluation of implemented, targeted, research development programs and services. As such, the NORDP Consultants Program has an unstated mission to contribute to the literature by documenting the impact of research development within the partner institutions that participate in the program. This literature and early results are documented in this chapter.

Case Study Narrative

Origin of the NORDP Consultants Program

In 2019, the representatives from the White House Initiative on HBCUs and the White House's Office of Science and Technology Policy engaged NORDP to discuss strategies for growing research capacity and competitiveness within HBCUs. Desired strategies included: assisting HBCU faculty with developing long-term plans for individual research programs and career trajectories, building sustainable expertise in ideation, identifying appropriate funding sources, engaging with funding organizations, developing competitive grant proposals, building and working in a team, collaborating with other institutions, and mentoring trainees (e.g., postdoctoral fellows, and graduate and undergraduate students).

These discussions led a small group of NORDP thought leaders to submit a proposal to the Schmidt Family Foundation venture, Schmidt Futures. NORDP proposed that four research development consultants would work with four HBCUs to provide research development services to strengthen their research capacity. To assess the effectiveness of different engagement modalities, two of those engagements would be primarily virtual and two would occur primarily in-person. NORDP was notified that the proposal was awarded in February 2020 but was delayed in launching the new project because of the Covid-19 pandemic that significantly disrupted the business of IHEs beginning in March 2020. Program implementation restarted in 2021.

Early Evolution: From Targeted Project to National Program

From submission of the initial proposal in 2019 to its earnest implementation in 2021, the world changed due to the Covid-19 pandemic. IHEs core businesses of education and discovery were disrupted, budgets were restricted, and many now-remote workers were reluctant to fully return to the office. These shifts brought changes to the implementation plan as well.

- *Change in principal investigator*—The original PI for the project became unavailable to implement the project after funding was secured. The NORDP Board of Directors explored the opportunity with several potential leaders for the project, including nesting it within existing committees, but ultimately appointed me—executive officer Kimberly Eck who was transitioning from the NORDP president to the immediate past president—to serve as the PI.
- *Re-positioning as a pilot project*—I quickly repositioned the Schmidt-funded project as a pilot for a larger national program. At the time the initial funding was secured, it was conceived of as a single two-year project. Although unknown at the time, I assumed there would be opportunities and interest to offer research development support beyond the four HBCUs partnering in the pilot project. Wanting to leave room for growth, I announced this larger vision as the NORDP Consultants Program and a founding mission that expanded to include MSIs and ERIs.
- *Programmatic changes*—I also made several programmatic changes to the pilot project. First, the rigid requirements that two engagements be conducted virtually and two in-person were relaxed. Each HBCU partner was free to plan virtual and in-person activities as needed. The world had thoroughly explored the advantages and disadvantages of remote work; whether consulting services could be offered virtually was already confirmed as a feasible approach. Second, consultants were assigned to teams of two to four and that was matched with an HBCU, which is a departure from the one-to-one pairings originally proposed. This teaming approach helped the program overcome two key logistic challenges. Busy research development professionals expressed hesitation at the 300 hours per year commitment. Spreading 300 hours across a team of consultants was much more palatable. Additionally, few consultants have all the deep expertise needed to support a complex IHE. Consultants also benefit from teaming with each other when they can share expertise.
- *External funding only*—Before launching the program, I decided that the program would operate exclusively using sponsored and philanthropic funding (e.g., from federal agencies and nonprofit foundations). An ERI

could not use its institutional funds to buy the NORDP Consultants Program. Although this eliminated a source of funding for the program, it allowed the program to remain focused on serving IHEs with the greatest needs and alleviated potential conflicts with NORDP members who work full-time as research development consultants. There would be no grappling over funds between the program and individuals.

- *Leveraging IHE infrastructure*—As an all-volunteer organization, NORDP did not have the infrastructure to administer federal funding. When it came time for the NORDP Consultants Program to apply for its first federal award, NORDP used the IHE infrastructure of its members rather than building its own. Emory University was the natural choice to administer the federal funding because it is my primary employer. This saved a significant amount of money and time for NORDP but also meant that Emory University, a private R1 institution with $1 billion in total research expenditures, became the administrative seat for the program. The irony that a large R1 institution receives funding to administer a program that supports MSIs and ERIs has been observed.

The pilot project and other early engagements are discussed in more detail below.

About the NORDP Consultants Program Today

The NORDP Consultants Program is dedicated to expanding our national research ecosystem by providing research development services to MSIs and ERIs at no cost to the institution. It is the only national program that aims to increase research capacity through the direct provision of research development services. To date, it has supported 172 ERIs from forty-four states and territories through three models of engagement described below. This included 73 MSIs, including 33 HBCUs, 26 HSIs, and 14 other IHEs with Asian American and Native American Pacific Islander-serving institution (AANAPISI), Alaska Native and Native Hawaiian-serving institution (ANNH), predominantly Black institution (PBI), and Native American-serving nontribal institution (NASNTI) designations.

Foundational Frameworks

Two foundational frameworks co-emerged simultaneously to program implementation. The right-sized research capacity framework and the socioecological approach to building research capacity. Given the relative dearth of literature about the research ecosystem, these frameworks were developed to better explain the program's approach to tailoring research development supports to meet an IHE's unique goals and needs as well as to illustrate how research development often works simultaneously across different levels of the research ecosystem.

Right-sized Research Capacity Framework

The core principle of the right-sized research capacity framework is that the number and type of resources and services in an institution's research enterprise should address the institution's needs in pursuit of its goals for the research enterprise. This framework acknowledges that many MSIs and ERIs do not seek to become R1 institutions. For example, many primarily undergraduate institutions may want to grow their research without offering graduate degree programs (a requirement for R1 classification). Simply overlaying R1 resources into an ERI without understanding the goals of the ERI is unlikely to be successful. This concept is infused into the approach used by the NORDP Consultants Program (Hemming and Eck 2024a).

Socioecological Approach to Building Research Capacity

The socioecological approach to the research ecosystem says:

1. Research capacity can be increased at the individual, department, school, institution, and regional levels and beyond.
2. Research capacity at any level can be influenced positively and negatively by policies, training, programs, activities, incentives, disincentives, resources, and other factors.
3. There is an interplay between the concentric circles. Capacity built in an inner circle has the potential to impact itself and the adjacent outer circle. However, capacity built in an outer circle has the potential to impact itself and all inner circles.

This framework helps the program navigate complex ecosystems and supports the evaluation team in assessing impact at multiple levels by actively considering the needs of each level (Hemming and Eck 2024b).

For example, on the individual level, the program might build research capacity by providing training in grant seeking, introducing faculty to program officers or arranging for them to visit campus, providing one-on-one proposal development coaching to faculty, and assisting in strengthening a proposal by providing review. At the department or school level, research capacity might be built by developing a course release policy, streamlining the approval process for research expenditures, or identifying best practices being implemented by peers to incentivize research in teaching intensive environments. At the institutional level, the program might support the institution in selecting and implementing its first electronic grants management system, develop a policy to redistribute facilities and administrative costs more strategically, or update policies and procedures to better align with recent research security requirements.

Personnel

The personnel involved in the NORDP Consultants Program grew and formalized over time. There are several categories of personnel that support the program, including consultants, peer coaches, budget doctors, the advisory group, project administration, and program leadership who are described below.

There are twenty-six active NORDP consultants and one alumna (an approved consultant who is no longer active in the program). Consultants are identified through an annual call for applications from potential candidates. All NORDP members with the requisite experience are eligible to apply. Consultants are expected to have broad expertise in research development and frequently bring experiences from research administration and other areas of the research infrastructure, as well as have demonstrated a commitment to the mission of the program by working with ERIs, MSIs, or faculty from historically excluded groups. Interested NORDP members submit a statement of interest that highlights their breadth of experience and commitment to the program mission. Applications are reviewed by a committee of peers (existing NORDP consultants) who make recommendations to program leadership about whether to approve an individual as a program consultant. Reference checks are conducted and invitations to join the pool of consultants are extended.

Once in the approved pool, opportunities to take on assignments (described in the next section) are presented to the consultants. Interested consultants are matched with MSIs and ERIs as needed. This process allows consultants to maintain control of their own schedule. Consultants provide research development and some research administration support to MSIs and ERIs across all types of engagements. All consultants are paid a uniform hourly rate, which streamlines contracting and negotiations. Occasionally, a sponsor pays a flat rate per proposal supported.

There are four peer coaches active in the program. Generally, peer coaches are former participants in the program. They provide one-on-one guidance to the MSI lead point of contact in the intensive cohort program (in contrast to those who support the growth of an IHE's research capacity). Peer coaches receive a stipend for this work.

There are four budget doctors that support the program. This nickname arose from their collective ability to diagnose and meet a participant's budget and submission preparation needs quickly. These experienced research administrators provide pre-award and post-award support to MSIs and ERIs during the embedded proposal development engagements. They receive a stipend for this work.

The advisory group is composed of four volunteers that represent the NORDP Board of Directors and previous program participants. The advisory group identifies high-level risks to the program and contributes thought leadership to mitigate those risks. The staff that lead the program includes a senior program director, a project director, and a program manager who are collectively responsible for directing and growing the program as well as implementing current projects.

Program leadership and administration includes myself, a full-time project director with oversight over the cohort models of engagement who is supported by a full-time project manager, and part-time director who oversees the short-term engagements.

Consultants and others are prompted to disclose outside employment in accordance with the policies of their primary employer, if applicable. Their contracts include clauses related to maintaining confidentiality and avoiding conflicts of interest.

Models of Engagement

The NORDP Consultants Program uses three models of engagement: intensive cohorts, embedded proposal development, and partner-initiated proposals. The initial pilot program used the intensive cohort format. The embedded proposal development and partner-initiated proposal models emerged soon after in direct response to invitations to collaborate from sponsors and ERI and MSI partners. Inputs, sponsors, number of MSI and ERIs engaged, and major activities for each model of engagement are presented in this section. Outcomes are presented in a subsequent section.

Intensive Cohorts

The intensive cohort engagements support MSIs in achieving their goals for growing research at their institution. A lead contact, often a leader within the research infrastructure, from each MSI is identified to serve as the primary contact for the NORDP consultants. Consultants use a three-phase (intake, planning, and implementation) engagement model with the MSI to iteratively create and implement a set of activities customized to the institution and documented in an engagement plan (template described in a later section). These engagements are summarized in Table 7.1 and additional details about the pilot project (Cohort 1) are presented in Eck (2023) and Eck et al. (unpublished[2]).

Embedded Proposal Development

In this model of engagement, the NORDP consultants provide proposal development support to all applicants invited to apply to a single federal funding competition. In these instances, the sponsor publicly invites potential applicants to submit a short (three to four pages) preliminary proposal using a typical announcement mechanism, such as a Dear Colleague letter or solicitation, and convenes a panel to invite potential applicants to participate in an innovative proposal development process with NORDP consultants. Consultants provide support through many phases of proposal development, including conceptualization, teaming (forming a consortium), budget development, calendaring, proposal production, and submission. The program provided embedded proposal development support in the NSF's Enabling Partnerships to Increase Innovation Capacity (EPIIC) and Advancing Research Capacity at HBCUs (ARC HBCU) programs. These engagements are summarized in Table 7.1 and more details about EPIIC are presented in Eck et al. (2023).

Table 7.1 Summary of Engagements Initiated 2022–2024

Model	*Engagement*	*Sponsor*	*Participating IHEs*	*Core Resources*
Intensive cohort	Pilot project[a] (Cohort 1)	Schmidt Family Foundation	4 HBCUs	300 hours per year provided by a team of 2–4 consultants + $15,000 infrastructure investment
	Cohort 2[b]	NSF OIA-2331578	4 MSIs	300 hours per year provided by a team of 2–4 consultants + $150,000 infrastructure investment and wrap-around supports
	Cohort 3[b]	NSF OIA-2331578	4 MSIs	300 hours per year provided by a team of 2–4 consultants + $150,000 infrastructure investment and wrap-around supports
	Cohort 4[d]	NSF OIA-2331578	4 MSIs (to be recruited)	300 hours per year provided by a team of 2–4 consultants + $150,000 infrastructure investment and wrap-around supports
	Cohort 5[d]	NSF OIA-2331578	4 MSIs (to be recruited)	300 hours per year provided by a team of 2–4 consultants + $150,000 infrastructure investment and wrap-around supports
Embedded proposal development support	EPIIC 1[a]	NSF ITE-2303003	49 ERIs	1 consultant supported each consortia application (~40 hours of support) + a budget doctor
	EPIIC 2[ac]	NSF ITE-2350104	50 ERIs	1 consultant supported each consortia application (~40 hours of support) + a budget doctor
	EPIIC 3[b]	NSF ITE-2350104	49 ERIs	1 consultant supported each consortia application (~40 hours of support) + a budget doctor
	ARC HBCU[ac]	NSF OIA-2337236	24 HBCUs	1 consultant supported each consortia application (~40 hours of support) + a budget doctor
Partner-initiated	Delaware State University[b]	NIH UC2GM157743	1 HBCU	450 hours per year provided by a team of 2–4 consultants

Source: NORDP Consultants Program activity data.

[a]Engagement activity is complete.

[b]Engagement is active.

[c]Evaluation is underway, including award notification and publication of results.

[d]Engagement planned for future years.

Partner-initiated Proposals

The NORDP Consultants Program also welcomes requests from MSIs and ERIs to participate in their proposals. These requests to collaborate are evaluated individually. If the request aligns with the mission of the NORDP Consultants Program and reasonable resources are available, the NORDP consultants develop a custom scope of work to support the MSI or ERI in achieving their goals for the proposal and project. One partner-initiated proposal has been awarded and others are pending or currently pending review. The NORDP consultants' work in each of the proposals focuses on conducting an institutional needs assessment for research capacity, creating an institutional action plan, and beginning to implement a high-priority items plan. See Table 7.1 for additional information on the awarded proposal.

Summary of Engagements to Date

In three short years, the NORDP Consultants Program has supported 172 ERIs, 73 of which are MSIs (including 33 HBCUs, 26 HSIs, and 14 other IHEs with AANAPISI, ANNH, PBI, and NASNTI designations) from 44 states and territories. Of the institutions that report research expenditures to HERD, their median total research expenditure is less than $3 million. The engagements below were initiated between January 2022 and December 2024. Only Cohort 1 and EPIIC 1 are complete with a final evaluation and publication. Other engagements are either ongoing, the evaluation is not yet finalized, or a detailed publication is under development. The program will recruit eight more MSIs into the intensive cohort model in future years and continue to engage in embedded proposal development program and partner-initiated proposals as opportunities arise. Currently, the program has funding for activities extending through 2028. Continuing the program beyond then will require securing additional external funding.

Table 7.1 presents a summary of engagements for which external funding has been committed. Each model of engagement has one or more engagements associated with it. The sponsors, number of participating IHEs (some IHEs participate in multiple engagements), and core resources for each are listed. A footnote indicates the stage of each engagement.

Tools, Templates, and Resources

The NORDP Consultants Program developed several tools, templates, and resources that have been used to further the goals of each engagement.

Three-phase Engagement Framework

For the intensive cohort engagements and partner-initiated proposals submitted to date, the NORDP consultants utilize a three-phase engagement framework, including an engagement plan template, that allows for flexibility and customization to ensure that the IHE partner receives appropriately tailored support while still advancing their shared programmatic goals. Unlike the earlier frameworks, which guide the overarching program philosophy, this framework serves to organize the activity of longer-term engagements.

Phase 1: Intake occurs during months one through three. In Phase 1, consultants collect and review institutional strategic plans, proposal and award data, information on centers and institutes, recent assessments or faculty feedback, and similar documents. Consultants also interview key stakeholders—such as vice presidents for research and staff, deans, associate deans for research, and successful researchers—to hone goals for the research enterprise as well as identify barriers, successes, and existing resources.

Phase 2: Planning occurs during months four and five. In Phase 2, consultants summarize information received, suggest potential activities, and identify potential areas for improvement. Using the engagement plan template (described in the next section), consultants document institutional goals for advancing the research enterprise as well as activities to achieve these goals.

Phase 3: Implementation occurs during months 6 to 24. In Phase 3, consultants implement the activities included in the engagement plan template and make revisions to the plan as needed. The engagement plan objectives and activities from Cohort 1 are available in Eck et al. (unpublished).

Engagement Plan Template

For the intensive cohort engagements and partner-initiated proposals submitted to date, NORDP consultants iteratively develop and use an engagement plan to guide their multi-year interaction with the MSI or ERI partner. The core of the engagement plan template prompts consultants to define SMART (*s*pecific, *m*easurable, *a*chievable, *r*ealistic, *t*ime-bound) objectives, metrics, milestones, activities, and other information as listed in Table 7.1. Critically, plans for sustainability are included in every single objective to ensure that capacity assessments and other activities are planned and documented in thoughtful ways so that the partner can continue the recommendations and activities after the engagement concludes.

The template includes the following elements and instructions to assist consultants creating their plan and is available in Eck (2023) with examples in Eck et al. (unpublished). For each SMART objective, the engagement plan template includes: one to three metrics that will demonstrate progress of each SMART objective, a baseline for those metrics, annual milestones, major activities that will be conducted to achieve the objective, a sustainability plan, and other contextual information needed to understand how the objective will be achieved. Though not paradigm shifting, this engagement plan template serves as an important touchstone throughout each multi-year engagement to align the expectations and actions of consultants and partners.

EPIIC Toolkit

The EPIIC toolkit—a resource for EPIIC applicants working on their full proposal alongside their NORDP consultant—includes explanations, examples, and sample language that consortia could use in the proposal drafting process. In EPIIC 2 and EPIIC 3, examples of successful proposals were added to this toolkit. Specifically, the toolkit included a submission checklist, project description outline, overview and rationale example, individual institutional plan templates, capacity-building examples, guided evaluation menu, broader impact examples, collaborative plan examples, data management plan template, collaborative proposal instructions for submission, and previous successful proposals.

The EPIIC toolkit is the first time a comprehensive, program-specific, proposal development resource suite has been available to all applicants competing in a federal funding program.

Equipping Minoritized and Emerging Research Institutions to Grow Their Enterprises (EMERGE) Library

The Equipping Minoritized and Emerging Research Institutions to Grow their Enterprises (EMERGE) Library[3] features co-created and peer-reviewed knowledge articles and case studies that address MSI and ERI research enterprise priorities and provides context-specific implementation strategies and insights.

It is the first national research resource library that centers the voices and experiences of MSIs and ERIs.

To encourage the development of EMERGE Library resources, the NORDP Consultants Program hosts writing sprints that bring together research development experts from MSIs, ERIs, and more resourced institutions to produce these resources. This innovative sprint model leverages and elevates the expertise of NORDP members and supports the coauthors, who have significant knowledge but limited incentive to publish. During a sprint, individual authors are recruited, teams are suggested, a template is provided, a topic is assigned to the suggested team, and the authors produce drafts that are peer-reviewed following a shared timeline. The first sprint resulted in the creation of three new peer-reviewed resources, each of which addresses a popular funding mechanism: the National Institutes of Health R16, National Institutes of Health R15, and the NSF CAREER.

Emerging Research Institution Assistance with Proposal Preparation to Leverage Ingenuity (APPLI) Guidebook

As a part of the NORDP Consultants Program's continued work with the NSF EPIIC program, a replication toolkit is being developed that is intended to be used by other funders interested in using innovative funding models and supports to increase research and innovation capacity within MSIs and ERIs. When finalized, the ERI Assistance with Proposal Preparation to Leverage Ingenuity (APPLI) Guidebook will be available in the EMERGE Library.

Data Collection and Analysis

The Cohort 1 and EPIIC 1 engagements have been completed and evaluated. The data collection and analysis procedures for each of these engagements are unique.

Cohort 1 with Four HBCUs

As described in Eck et al. (unpublished), the external evaluators used information collected through several methods during the second year of the program. The evaluators also conducted thirty-minute interviews with consultants and HBCU representatives at the midpoint of the program in

March 2023. The purpose of these interviews was to collect in-depth, qualitative information on consultant interactions with their partnering HBCU; progress made toward improving research development and administration within each institution; and engagement plans for the last year of the program. The evaluators also conducted thirty-minute interviews with consultants and HBCU representatives at the end of the program in December 2023. Interview questions assessed the overall changes made to the HBCUs research development and administration processes as well as the achievement of goals set in the beginning of the project. The evaluators obtained program-tracking data from each consultant team. Tracking data included information on NORDP consultant-led activities, research grants submitted and awarded, research development documents reviewed and created, and the number of connections made.

EPIIC 1 with Forty-nine ERIs

As described in Eck et al. (2023), after the forty-nine institutions completed the submission of their full proposal to the NSF EPIIC solicitation, the participants who attended the in-person workshop and submitted a proposal (n = 120) were asked to provide feedback about the resources provided by the NORDP Consultants Program. The online survey was administered by me using Qualtrics for the purpose of program evaluation and improving the process in the future. Likert scales and open responses were used to capture respondent feedback. Sixty-six participants responded, yielding a response rate of 55 percent.

Findings

Cohort 1 with Four HBCUs

The four HBCUs participating in the pilot project (Cohort 1) made significant progress during their two-year engagement. Working with the NORDP consultants, each institution identified key challenges and created an engagement plan to address those challenges and strengthen their research capacity at multiple levels (individual, institutional, regional, and beyond). Three of four institutions achieved all the SMART objectives identified in their engagement plans. Details about Cohort 1 are reported in Eck et al. (unpublished). Below is a summary of the activities completed in Cohort 1.

At the first HBCU, NORDP consultants reviewed thirteen existing documents and produced two new policies and processes to build research operations efficacy and efficiency; provided technical assistance to a director within research administration to support institutional planning and research management activities; implemented sixty professional development activities to grow skills to compete for federal and private funding; delivered nine activities specifically related to identifying appropriate funding sources; and provided proposal development support for four proposals.

At the second HBCU, NORDP consultants developed a web-based research hub to increase the visibility of the institution's research operations, resources, and research expertise; hosted workshops on topics including funding agencies and sponsors, strategic planning for research careers, and partnerships and collaboration; co-designed a new-faculty orientation to highlight research; delivered a comprehensive research capacity report to institutional leadership that described strategies to enhance research development capabilities; increased support from institutional leadership to promote initiatives that strengthen the institution's research culture; identified long-term sustainable strategies to grow grant-seeking efforts; facilitated networking within NORDP to increase access to examples of and technical assistance for aspects of the research enterprise; produced 10 research development documents, identified and began implementing research administration software; provided professional development and training to expand the knowledge of research development functions; established two organizational partnerships and facilitated connections with ten other IHEs; and provided small stipends to incentivize faculty to work one-on-one with NORDP consultants to develop and submit an external proposal.

At the third HBCU, NORDP consultants renegotiated the indirect cost rate with the US Department of Health and Human Services; identified and began to implement the institution's first electronic research administration system; provided one-on-one support for six proposals; reviewed ninety-four existing documents and produced twenty-eight new research development documents; developed a new faculty incentive package to increase overall faculty engagement in research; supported faculty in building science communication skills to raise their faculty profile and visibility; supported faculty ideation and developed ten research projects; and created an undergraduate research program to increase student research participation.

At the fourth HBCU, NORDP consultants identified and began implementing an electronic research administration system to streamline processes; reviewed thirty-five existing documents and produced ten new research development documents; supported the expansion of the research development office by recruiting the inaugural director and other personnel; provided proposal development support for eighteen proposals; hosted seventy-three sessions with faculty and research administrative staff members to advance their research development knowledge, skills, and abilities; and provided seventy-fiveone-on-one faculty consultation services to support research career planning.

These intensive activities helped to strengthen the research capacity at the participating HBCUs at multiple levels and bolstered awards totaling $41.2 million. As summarized in Eck et al. (unpublished):

> In just two short years, the four participating HBCUs (Cohort 1) and their NORDP consultants made significant progress on growing research capacity, paving the way for future investments. Of the 20 objectives, four focused exclusively on building individual investigator's capacity to compete for federal and private funding. Eleven objectives pursued institutional capacity to improve research development and administration processes and integrate research into the larger enterprise; five worked across multiple intervention levels to identify appropriate funding sources and engage with funding organizations. While no institution defined a discrete objective to intervene at the regional level, the program design and the needs of each institution resulted in various forms of institutional networking resulting in over 21 new connections and 14 new partnerships.

EPIIC 1 with Forty-nine ERIs

During the in-person workshop and post-workshop periods, more than 95 percent of respondents indicated that a NORDP consultant helped them refine their project concept or identify potential collaborators. On average, 97 percent of respondents (n = 66) rated NORDP consultants as very good or excellent on responsiveness, reliability, listening skills, proposal development knowledge, and effective communication. The budget doctor and EPIIC toolkit received similarly positive ratings. When asked to identify the most useful piece of advice or resources provided by the NORDP consultants, the two most common responses focused on concept development (n = 15) and the consultants' knowledge of NSF and proposal development (n = 13). One individual offered the following reflection in an open-ended response:

> I recommend this model to become a standard in future preproposal initiatives. I have never experienced an agency taking a direct approach to assist HBCUs and MSIs to ensure our success. This demonstrates your commitment to our universities and we recognize and appreciate it. I applaud the agency for this engagement and hope to see more of it. It made a huge difference in our submittal, our respect for the agency and the collaboration and future partnerships with other universities will help us all in our efforts to build capacity in research, translation of research into practice and to develop a diverse workforce. Thank you, NSF and the NORDP consultants for all of your support and assistance.

After the NSF merit review, 13 of the 13 EPIIC consortia proposals were funded. All (100%) of the institutions that submitted a full proposal in response to the NSF EPIIC program were successful. In total, NSF awarded $19.6 million using this innovative funding model. Notably, with regard to expanding the research ecosystem, this cohort of EPIIC awardees include three ERIs that had never secured an NSF award as a prime awardee and two more IHEs that had not secured an NSF award as the prime awardee in more than twenty-five years. Additional details about EPIIC 1 are reported in Eck et al. (2023).

EPIIC 2 with Forty-eight ERIs

A detailed publication that summarizes EPIIC 2 and its evaluation is in development with coauthors. We do have the award results. After the NSF merit review, 14 of the 14 EPIIC consortia proposals were funded. Of the institutions that submitted a full proposal in response to EPIIC 2, 96 percent were successful. In total, the NSF awarded $18.6 million using this innovative funding model. Notably, with regard to expanding the research ecosystem, this cohort of EPIIC awardees included one ERI that had never before secured an NSF award as a prime awardee.

ARC HBCU with Twenty-four HBCUs

A detailed publication that summarizes ARC HBCU and its evaluation is under development with coauthors. We do have the award results. After the NSF merit review, four of the six ARC HBCU consortia proposals were funded in full, and the remaining two received planning grants. In total, the NSF awarded $10.5 million through this program.

Discussion and Reflection

Feasibility

In total to date, the NORDP Consultants Program has worked with 172 ERIs, 73 of which are MSIs (including 33 HBCUs, 26 HSIs, and 14 other IHEs with AANAPISI, ANNH, PBI, and NASNTI designations) from forty-four states and territories to secure an incredible $90.6 million, including $63.4 million to invest in continued research capacity-building activities at, and resources for, their institutions. Individuals from the IHEs that have engaged with the program have also benefited by building their individual research capacity by increasing the skills necessary for grant seeking, proposal development, and research administration through training, one-on-one coaching, and proposal development support.

The NORDP Consultants Program demonstrated the feasibility of providing research development support to MSIs and ERIs using a consultant model and documented the multi-level impact that research development can have on IHEs. The program fills a gap in the research ecosystem by serving institutions that do not have research development, have limited research development support, and do not have the resources to engage for-profit consultants on a long-term basis. Further, using experienced consultants vetted through a peer review process and relying on external funding alleviates the burden of networking, vetting, and managing a bid process for consulting services by MSIs and ERIs.

The program has demonstrated that research development support and critical infusions of research administration advice can be offered effectively through consultants as opposed to permanent staff. This lends itself to a more sustainable model for emerging research operations that are not yet able to support full-time research development staff by accelerating a cycle of external funding for the IHE and supporting the IHE in securing additional funds for investing in research capacity. If NORDP consultants can help ERIs secure more external funding in a short period of time, the IHEs will recover more facilities and administrative costs and have more resources available to make long-term investments and eventually hire permanent staff. In other words, although not a panacea, consultants can provide a jumpstart to the research ecosystem at an ERI.

A Role for Professional Societies

The NORDP Consultants Program also demonstrated the role for professional societies in strengthening and expanding the national research ecosystem. Most NORDP members are employed at a single IHE, institutions that frequently compete against each other; yet the organization has cultivated a supportive, volunteer-driven, collaborative environment. The NORDP Consultants Program effectively leveraged the NORDP network and expertise of its community.

Looking Ahead

The NORDP Consultants Program will continue to operate for the duration of its funded projects, guided by the program leadership and advisory group. The program will make multiple contributions to the literature through the EMERGE Library and articles that describe the impact of research development on the participating IHEs.

Conclusions

Professional organizations, like NORDP, can have a unique role in strengthening and expanding the research development ecosystem that is complementary to for-profit consulting businesses. Programs like the NORDP Consultants Program that use a consulting model can jumpstart the research ecosystem at an MSI or ERI, which can lead to a cycle of increasing funding and investment in the research infrastructure. Tapping into the network of a professional organization allows MSIs and ERIs to benefit from consultants with rich expertise with whom they may not have otherwise connected. This program has helped MSIs and ERIs to secure significant external funding and is helping to create a more robust national STEM community by securing additional funding to invest in research capacity, directly providing research development support to strengthen IHEs infrastructure, and in some cases, supporting participating IHEs in securing their first NSF award.

Notes

1 The NORDP Consultants Program is not possible without the consultants, peer coaches, budget doctors, advisory group members, program staff, and MSIs and ERIs that engage in the program. Although summarized by a single author in this book chapter, dozens of others have contributed through leadership to the program during its initial conceptualization and throughout its evolution. Partners from the NSF, Advancing Research Impacts in Society, and Know Innovation have helped the program grow and evolve in unanticipated and amazing ways. The National Organization of Research Development Professionals and Emory University have been generous with their support for the program.

2 This manuscript, entitled "Using Research Development to Enhance Research Capacity in Four HBCUs: Results from the NORDP Consultants Program Cohort 1," has been accepted for publication at *Research Management Review.*

3 The EMERGE Resource Library is available at https://researchdevelopmentconsultants.emory.edu/resources/index.html.

References

Eck, K. 2023. Feasibility and Early Results of Providing Research Development Consulting Services to Historically Black Colleges and Universities (HBCUs): Year 1 of the NORDP Consultants Pilot Project. National Organization of Research Development Professionals. https://nordp.memberclicks.net/assets/NORD/Eck_Feasibility%20and%20Early%20Results_FINAL.pdf.

Eck, K., and J. Roney. 2022. What Is Research Development? National Organization of Research Development Professionals. https://nordp.memberclicks.net/assets/NORD/Eck_Roney_WhatIsRD_.pdf.

Eck, K., S. Renoe, and A. Pai. 2023. *Real-Time Capacity Building: Feasibility of Embedding Research Development Consultants in a Federal Funding Competition.* National Organization of Research Development Professionals. https://nordp.memberclicks.net/assets/NORD/Eck_Renoe_Pai_2024_Article.pdf.

Eck, K., K. Eck, M. Fechner, R. Latimer, and K. Tindle. 2020. NORDP Annual Salary Survey—2020. National Organization of Research Development Professionals.

Hemming, J., and K. Eck. 2024a. "A Culturally Relevant Praxis: Applying Participatory Action Research Principles to Support MSIs in Growing Research Capacity." Poster presented at the Sixteenth Annual NORDP Research Development Conference, Bellevue, Washington, April 28–May 1, 2024. https://doi.org/10.5281/zenodo.13357659.

Hemming, J., and K. Eck. 2024b. "A Socio-ecological View of Research Capacity." PowerPoint presentation at the First Annual NORDP Research Development Consultants Program Annual Meeting, Miami, Florida, September 13, 2024. https://doi.org/10.5281/zenodo.13756004.

Hofstra, B., V. V. Kulkarni, S. Munoz-Najar Galvez, B. He, D. Jurafsky, and D. A. McFarland. 2020. "The Diversity–Innovation Paradox in Science." *Proceedings of the National Academy of Sciences of the United States of America* 117 (17): 9284–91. https://doi.org/10.1073/pnas.1915378117.

NCSES (National Center for Science and Engineering Statistics). 2022. "Higher Education Research and Development Survey Table Builder." National Science Foundation. https://ncsesdata.nsf.gov/builder/herd.

Pai, A., K. Eck, S. Renoe, J. Brown Clarke, B. Brown, C. Coley, M. Collier- Youngblood, H. Fields, L. Hammonds-Odie, T. Hendrickson, G. Hollowell, D. Huizinga, T. Inniss, A. Maglia, R. Nader, C. Talley, S. Vassmer, T. Washington, J. Weintraub, K. Williams, and L. Winfield. 2024. *Strategies for Building Capacity at Minority Serving Institutions for Advancing Research and Research Impacts*. Center for Advancing Research Impact in Society. https://doi.org/10.32469/10355/98061.

Preuss, M., K. Eck, M. Fechner, and L. Walker. 2018. "Describing Research Development: A First Step." *Research Management Review* 23 (1): 1–19. https://files.eric.ed.gov/fulltext/EJ1215447.pdf.

Preuss, M., K. Eck, M. Fechner, and L. Walker. 2020. "Research Development and Its Workforce: An Evidence-Based Compendium for Higher Education and Other Environments." *International Journal on Studies in Education* 2 (1): 1–25. https://ijonse.net/index.php/ijonse/article/view/12.

Quider, A., and G. Blazey. 2023. "How to Keep Emerging Research Institutions from Slipping through the Cracks." *Issues in Science and Technology* 39 (3): 50–3. https://doi.org/10.58875/ACQP2387.

8

The Society of Women Engineers: Promoting Transparency to Foster a Culture of Belonging

Roberta Rincon and Nicola Mohan

Professional STEM societies play a pivotal role in changing the engineering profession to better reflect the diverse populations that they serve.[1] Many of these societies span across education and career, offering students and professionals from various sectors and industries opportunities to engage with others in their technical disciplines. For STEM societies with national or international scope, members interact with people of different backgrounds, geographies, and experiences. This is professionally beneficial for members, but it also presents societies with a challenge and responsibility to advance DEI within the profession so that all members feel a sense of belonging both within the organization and across the STEM workforce.

A recent review of how professional STEM societies can integrate diversity, equity, inclusion, justice, and accessibility into their functions noted that not only are these societies well-positioned to do this work but their members consider the organizations to be very important for catalyzing change (Meyer-Gutbrod et al. 2023). Funders have seen the role that professional STEM societies can serve and offer funding mechanisms to support diversity-focused cultural change, including the NSF (Campbell-Montalvo et al. 2022b). However, culture change is difficult and requires addressing structural factors that define behavioral expectations, policies, and access to resources as well as social factors that allow biases, stereotypes, and microaggressions to negatively impact organizational climate (NASEM 2021). In addition, implementing strategies to improve diversity, equity, inclusion, and belonging (DEI&B) in a professional STEM society requires reflective and intentional effort (Madzima and MacIntosh 2021). Organizations that are effective in this approach typically promote greater diversity, representation, and inclusion; establish trusted reporting systems; and build transparency and accountability into the intervention process (Shiffman et al. 2022).

This case study focuses on the experiences of the Society of Women Engineers (SWE), a professional STEM organization whose mission is to empower women to achieve their full potential in careers as engineers and leaders. SWE has long taken a strong stance for the inclusion of women from all backgrounds in the engineering and technology professions. The organization has sought to be a leader in this space by incorporating DEI&B as a strategic priority and establishing goals and targets for success. Through this analysis of SWE's progress in addressing this strategic focus, we seek to provide insights into how other professional STEM societies can effectively measure progress toward achieving their own DEI&B goals.

Background

SWE was founded in 1950, when less than 1 percent of working engineers were women (Corbett and Hill 2015). Archival records from that period highlight the challenges that women experienced at the time, which ranged from broad societal attitudes about acceptable gender roles to the day-to-day challenges of being one of few, such as a lack of women's restrooms where engineers study and work (SWE 2020). SWE was established to give women engineers a community and a voice within the engineering industry.

Since its beginning SWE leaders have focused on ensuring that the society is seen as a professional organization within engineering. Membership was open to anyone regardless of race, as organizers were more focused on supporting women who had done the work required to become engineers and less on the social constructs that limited women's inclusion in the profession. SWE's purpose as stated in its original articles of incorporation was "to inform the public of the qualifications, abilities, and achievements of women engineers to bring about the better utilization of engineering talent for the economic and social benefit of humanity and to encourage young women with suitable aptitudes and interest to enter the engineering profession and to guide them in their educational programs" (SWE 1952). It would be decades before any records were kept on SWE members that noted such demographic information as race or ethnicity (Eller English 2023).

In the mid-1980s SWE developed its first mission statement. The statement aligned with the organization's purpose, which focused on helping women to "achieve their full potential in careers as engineers and leaders," promoting engineering "as a positive force in improving the quality of life," and demonstrating "the value of diversity" (Eller English 2023). This was the first introduction of diversity as a core value to the organization.

Over the years, SWE has continued its journey to develop a more diverse and inclusive membership and engineering community. In 1998 the organization adopted its first diversity principles, which included race/ethnicity, age, physical abilities, and sexual orientation but did not include religion or gender (Eller English 2023). In 2013 the diversity principles were expanded to include "sexual and affectional orientation, actual or perceived gender, gender identity and expression" (Eller English 2023). Starting in 2018, SWE began including diversity and inclusion in its long-term strategic plans, and in 2020 the organization introduced the term "DEI" to highlight the necessity of equity alongside diversity and inclusion. For SWE, equity means treating all individuals fairly and justly while acknowledging that different individuals may need varying levels of support and resources to have equal opportunities for success (George et al. 2023). Its addition to diversity and inclusion signified a progressive growth in understanding that true inclusion requires actively dismantling systemic barriers and providing equitable opportunities for everyone. In 2021 SWE adopted a new DEI statement that recognized members' intersecting identities and acknowledged gender as a continuum, stating that "everyone is welcome to join SWE as a member" (Doty 2021). Belonging was added the following year, which made DEI&B a major focus for the organization. In essence, the society aims to be a space where women in engineering and technology can be their authentic selves, free from the pressure to dim their light to fit in.

As of 2025 SWE has more than 50,000 members and more than 500 sections and affiliates around the globe. This membership is somewhat evenly split between collegiate and professional members, spanning college to career. While the society hosts several national and regional events to bring members together, much of the engagement occurs at a local level within the sections and affiliates. In addition to sections and affiliates, SWE has twenty-two affinity groups to bring together members with similar interests and backgrounds. The SWE affinity groups are organized into Diversity, Equity, Inclusion, Belonging Affinity Groups; Business and Interest Affinity Groups; and Career Stages Affinity Groups.

Including DEI&B as a SWE strategic priority brings greater accountability and transparency by requiring clear measures of progress toward stated goals. This chapter examines SWE's efforts to create a more diverse, inclusive, and equitable community, both within the organization and across the engineering sector. Using publicly available documentation—including SWE publications and archived materials as well as data collected from self-assessments and interviews with SWE staff involved with these efforts—this case study considers the methods SWE has taken to measure progress toward its DEI&B goals. Professional societies and similar organizations that seek to incorporate DEI&B as a core value and strategic priority can learn from the challenges and successes that SWE has experienced during this journey.

Theoretical Framework

Culture change requires significant communication, transparency, and accountability to be successful. Research has shown that interventions to improve diversity and inclusion within an organization are not a one-and-done effort; rather, they take time, patience, and significant support from leadership (Bezrukova et al. 2016; Chavez and Weisinger 2008). Investigating and discovering the factors that contribute to a successful or failed implementation should consider the execution methods and the evaluation process used to measure progress.

This study applies a theory of change framework to understand the iterative process that SWE has undertaken to achieve its DEI&B goals. Theory of change is often applied to interventions that have short- and long-term outcomes, which allows practitioners to test hypotheses and assumptions as actions and strategies are implemented (Taplin and Clark 2012). The theory of change framework is often used during planning and implementation, but it is extremely useful when measuring progress and communicating changes to affected audiences (Gienapp and Hostetter 2022).

In addition to analyzing the goals, strategies, and feedback obtained from the data, we used reflection questions that aligned with the theory of change framework to guide the case study analysis and help us understand how SWE measures its progress toward its DEI&B goals. Reflecting on the progress made since DEI&B became a strategic focus of the organization included considering the assumptions that leadership held at the time regarding the need for and acceptance of culture change, the alternative future that they wanted to achieve, and the observable changes that have occurred.

Case Study Narrative

The shift from using the term "D&I" to "DEI" and finally to "DEI&B" represents a gradual evolution in SWE's approach to tackling diversity and inclusion issues. Each stage in this transition underscores an increasing understanding of the complexity and challenges involved in creating an inclusive environment.

Incorporating DEI&B into the Strategic Plan

Beginning in 2013, SWE aimed to acknowledge and celebrate diversity in every aspect of the organization. Concurrently, SWE emphasized fostering an inclusive culture where every individual in the SWE community felt welcomed and respected regardless of their diverse backgrounds. However, it was not until 2018 that the organization formally incorporated D&I into its multi-year strategic plan with specific goals and metrics.

At the time, leadership believed the organization was not where they wanted it to be. This was more than an assumption; while the society's leadership was visibly lacking in diversity, the data also showed that SWE did not reflect the diversity that exists in the engineering profession. SWE members were becoming more vocal about the lack of diversity and inclusion within the society.

In developing SWE's strategic plan for fiscal year (FY) 2019, SWE Senate's Annual Strategic Planning Subteam conducted the necessary research to update the pillars included in the new plan. With the help of an external consulting group to guide the strategic planning process, the subteam made recommendations to the board of directors, who adopted the four-year strategic plan in 2018 to include D&I as a goal with associated strategic objectives.

The impetus for adding D&I to the strategic plan came directly from SWE members. Members from diverse backgrounds shared their experiences of feeling a lack of inclusion within the organization in one-on-one conversations, town halls, and member surveys. These experiences often occurred at the section level, which affected members' sense of belonging. This was extremely concerning for leadership, who believed that a diversity-serving organization should foster feelings of inclusion and belonging if its purpose is to support diverse engineering talent. Beyond this, the membership expected SWE to uphold its espoused diversity values. However, without incorporating these values into SWE's strategic goals, leadership knew that members would not know their concerns were being addressed and progress toward D&I goals would not be tracked.

SWE's strategic goals and objectives are publicly available and shared across the organization. Including D&I in the strategic plan highlighted that it was an organizational priority and set the expectation that the organization should reflect the diversity seen in the engineering profession. The strategic plan cascades down from the board of directors to committees, section leadership, and SWE headquarters staff, with detailed goals and assigned responsibilities. In FY 2023 SWE's board of directors created specific strategic objectives aligned with DEI&B, including to "foster an inclusive culture across the society so that all within the SWE community experience a sense of belonging." This is one objective that drives the SWE operational plan, which outlines initiatives to achieve it.

SWE's senate serves as the strategic body for the society. The senate works continuously to maintain forward progress toward SWE's strategic goals. This ongoing effort is demonstrated by the addition of equity and belonging in 2021 and 2022, respectively, in response to member feedback. It also reflects the recognition that advancing DEI&B is a continuous process.

The shift from D&I to DEI and then to DEI&B reflects SWE's continuous journey toward fostering a diverse, equitable, inclusive, and truly welcoming community for all its members. By embracing these principles, SWE not only strengthens the organization but also sets an example for the engineering profession and beyond, fostering positive societal change by championing the value of diversity. If SWE encourages its members to seek employers whose workforce reflects the diversity they wish to see, then SWE should reflect the diversity that it says is vital to a strong and competitive STEM workforce.

Setting Goals and Developing a Baseline

As an ongoing effort, SWE's strategic plan includes short- and long-term goals. When D&I was initially incorporated into the strategic plan, the short-term focus was on improving data collection and reporting on member demographics. At the time, SWE's demographic data collection occurred in various places, including member applications, event registrations, and surveys, and there was not a consistent method for requesting this data. The long-term goal of having the membership reflect the demographic makeup of the engineering profession required consistency in how demographic data was being collected, encouragement to members to provide the organization with this data, and the development of a baseline to begin tracking progress toward D&I goals. Another long-term goal, introduced as a strategic objective in FY 2023, is to ensure "multifaceted representation at all leadership levels through intentional efforts to diversify SWE's leadership pipeline and membership." Accurate demographic data at each level of the organizational leadership is critical to achieving this objective.

In 2019 fewer than 70 percent of member records included race/ethnicity. Even fewer were sharing with SWE their age, sexuality, or disability status. SWE was also uncertain of the reliability of their demographic data, which made it difficult for the organization to determine a baseline for D&I metrics. The following year, SWE collaborated with an external consultant to conduct a comprehensive analysis of its demographic data, including the questions asked, the timing of these questions, and the appropriate answer options. The objective was to enhance SWE's understanding of its membership and key programmatic points while aligning with its organizational values and forward-thinking goals related to its D&I efforts. Through this partnership, SWE collaboratively designed and implemented a streamlined process for data collection. This process ensured consistency and accuracy in gathering demographic data across various aspects of SWE's programs, including membership, conferences and events, awards programming, scholarships, and the leadership pipeline (George et al. 2023).

As a result of this work, SWE successfully standardized the collection of the following key data points across all programmatic areas for use in FY 2021:

- Gender identity
- Pronouns
- LGBTQ+ identity
- Race and ethnicity
- Language
- Age
- Level of education
- Technical discipline
- Education affiliation
- Country and state
- Employment status
- Years of experience
- Employer type
- Employer size

Additionally, SWE changed the demographic questions from optional to mandatory, requiring members to choose an option—even if it is "prefer not to answer." Despite this change, as of 2024, the number of member records missing a response for race/ethnicity remains high. When reviewing member records staff discovered that the member database includes long-standing members who have not updated their records. While this represents another challenge for the organization to address, SWE has improved consistency and developed a baseline to measure progress toward its DEI&B goals.

While tracking race/ethnicity was essential for ensuring that SWE could monitor the diversity of its leaders and members, much discussion focused on the importance of gender identity as a measure of inclusion. As a women-centered organization, SWE aims to stay true to its mission of supporting women while also being inclusive of other genders. In 2019 one in four members was missing gender information in their SWE member records. In 2021 when the board of directors adopted a new DEI statement, they explicitly welcomed nonbinary and gender-expansive individuals while still defining SWE as a women-oriented organization. To align with the new statement, staff considered ways to avoid the assumption of a gender binary when collecting member demographics and promoting the use of gender pronouns in publications, during events, and on name badges. By normalizing the practice of sharing pronouns, establishing the consistent collection of gender identity, and allowing people to self-describe their gender identity when they prefer to do so, SWE officially recognizes and honors all self-identified genders. Through these strategies SWE aims to foster a welcoming and inclusive environment where individuals of all gender identities are acknowledged, respected, and valued.

Implementing Strategies and Measuring Progress

SWE leadership tracks progress toward its DEI&B goals through a key performance indicator (KPI) dashboard. When D&I first became a strategic priority, the KPI dashboard provided quantitative data to track member diversity, society-level leader diversity, and member perceptions of D&I within SWE. When the strategic plan was renewed in 2022, the KPI dashboard incorporated section and affiliate leader diversity, belonging measures, and the diversity of award recipients. Some of these new metrics were included as a result of a DEI self-assessment pilot that SWE participated in with the American Association for the Advancement of Science in 2021, which involved reviewing SWE's diversity principles, policies, training requirements, and demographic data to encourage professional societies like SWE to make DEI the norm within their organizations. The exercise highlighted gaps at SWE, resulted in significant changes around the society's DEI&B metrics, and required training for SWE leaders and staff. For example, all SWE society leaders, section leaders, and staff members are now required to take a virtual DEI&B training course that includes multiple modules on preventing harassment and discrimination. Additionally, DEI&B moments are now standard practice across all SWE staff, committee, and board meetings of more than one hour, which highlights the priority of the society's espoused DEI&B values (George et al. 2023). The purpose of these changes is to ensure that those who are involved in implementing SWE's strategic priorities have a baseline understanding of what the organization is trying to accomplish. Additional DEI&B courses are offered through SWE's Advance Learning Center as a continuous learning opportunity for members, including several courses available in multiple languages.

The KPI dashboard includes multiple metrics and associated indicators that note the frequency of data collection, the timing within the year it is collected, and the data collection method. Standardizing data points is a lengthy process and demonstrates SWE's overall commitment to establishing a data framework that will allow the organization to make informed decisions (George et al. 2023). As a result of this work, SWE began consistently reporting on the existing KPI metrics in FY 2022. SWE also set new KPI metrics for the board of directors and senate aligned with the new four-year strategic plan developed that year. In FY 2023 SWE established the first baseline data sets for the KPI dashboard that aligned the current strategic plan with the new metrics from standardized language.

To complement the most recent strategic plan for 2023 to 2027, SWE introduced a set of KPIs in the aggregate and by key demographic subsets (ethnic identity, belonging in the LGBTQ+ community, gender identity, disability status, and global location). In the aggregate, SWE's member data indicated positive diversity trends within the organization; however, disaggregation by demographic subsets uncovered issues affecting member inclusion and belonging that were undetectable at a high level. Despite SWE's positive position when compared against other organizations' KPI scores, SWE leaders noted that feelings of belonging are best understood and addressed when analysis, accountability, and reporting are conducted at a deeper level.

The board of directors and senate reviewed the KPI metrics and shared the public-facing metrics through annual reports to members and the public. Table 8.1 provides a sample of the select data from the KPIs shared in the 2023 and 2024 DEI&B reports, including the baseline and target for each metric. The report provides a more detailed breakdown of results by various demographic groups, including race/ethnicity, sexual orientation, and disability status. By providing this data to members and the public, SWE aims to ensure that its assessment of progress and impact is communicated to all stakeholders and sets the expectation that this information will be shared regularly.

In addition to the quantitative measures, SWE implemented focus groups to gather qualitative data to understand member perceptions of inclusion and belonging. SWE leaders were particularly interested in hearing from members of underrepresented groups within the organization, who were easily overlooked in the KPI dashboard due to their low numbers. As part of the strategy to increase communication and feedback loops regarding key subsets of SWE members, SWE started conducting focus groups for key groups and all associated affinity groups on a three-year cycle. All focus group participants are compensated for their time and input, an important recognition of the valuable contributions of those most impacted by a lack of inclusion and belonging within the organization. During these focus groups, members provide insights into their perspectives on DEI&B at SWE. Table 8.2 lists the planned focus group calendar for FY 2023 to FY 2025.

SWE partners with external consultants to support its strategic planning process, including annual member surveys to gather insights into member perceptions of DEI&B and organizational culture. In addition to the annual surveys, SWE also assesses members' sense of belonging during its events. These surveys provide the board of directors with valuable data to monitor the progress toward SWE's DEI&B objectives, and the surveys have changed as the objectives and metrics have changed. The addition of "belonging" to the annual member survey has revealed variations in experiences between local and society levels and among collegiate and professional members.

Another significant strategy that influences SWE's progress toward its DEI&B goals involves marketing. Communication changes in the nomination process for leadership positions, awards, and recognitions are essential to ensure all members understand the procedures. In recruiting diverse candidates for leadership roles, awards, and scholarships, SWE recognized that traditional information-sharing methods about the nomination process were not effectively reaching the desired

Table 8.1 Select Data from the Society of Women Engineers' (SWE) Key Performance Indicator Metrics

Metrics and Society Health Indicators	**Measures Included**	**Baseline**[a]	**Fiscal Year Average**	
			2023	**2024**
Net promoter score	Based on survey question: How likely would you be to recommend SWE membership to a colleague or peer? (To assess word-of-mouth loyalty, a critical factor in association growth.)	36	46	53
Satisfaction score[b]	Overall, how satisfied are you with your membership so far?	4.13	4.27	4.31
Diversity perception score[b]	Please indicate your level of agreement or disagreement with the following statement: SWE is a diverse organization working to mirror the diversity of the industry.	3.84	4.12	3.84[c]
Member engagement score (index)	Average response score on a 6-point scale: 1. Plan to renew their membership. 2. Have accessed a SWE resource or participated in an event in the past month. 3. Consider SWE to be their primary membership society. 4. Have talked to a colleague about SWE in the past three months. 5. Have volunteered with SWE in the past or would like to do so in the future. 6. Have donated to SWE in the past 12 months.	4.03	4.27	4.28

Sources: George et al. (2023) and Mohan et al. (2024).

[a]Baseline set using data collected in 2019.

[b]Average response score on a 5-point scale.

[c]In 2024 two new measures were added to specifically measure belonging: (1) In your local area, please rate your sense of belonging to the SWE community from 1 to 10; (2) At the society level, please rate your sense of belonging to the SWE community from 1 to 10.

audiences. To address this, SWE needed to adopt more innovative approaches to promote leadership opportunities. Additionally, members from various backgrounds, especially those underrepresented in leadership, needed to see themselves reflected in these roles. For example, anecdotal evidence indicates that highly accomplished SWE members hesitate to nominate themselves for leadership positions due to self-doubts about their qualifications. In response,

Table 8.2 Planned Focus Groups for Fiscal Years 2023 to 2025

Fiscal Year	*Month and Year*	*Location*[a]	*Participants*
2023[b]	October 2022	WE22—Houston	Collegiate members
	October 2022	WE22—Houston	Affinity group members
	March 2023	Virtual	Neurodivergent and members with disabilities
	April 2023	Virtual	Asian American and Pacific Islander members
	May 2023	WELocal—Barcelona	Global regions
2024	July 2023	Virtual	Black and African American members
	October 2023	WE23—Los Angeles	Late-career and retired members
	October 2023	WE23—Los Angeles	Early career members
	February 2024	Virtual	LGBTQ+ and nonbinary members
	April 2024	Virtual	Latinx and Hispanic members
2025[c]	August 2024	Virtual	Indigenous members
	October 2024	WE24—Chicago	Midcareer members
	October 2024	WE24—Chicago	Affinity group members
	February 2025	Virtual	Multi-racial members
	April 2025	Virtual	New members
	Spring 2025	Virtual	Global regions

Sources: George et al. (2023) and Mohan et al. (2024).

[a]WE22, WE23, WE24, and WELocal refer to SWE's regional and annual conferences.

[b]July 1, 2022 to June 30, 2023.

[c]July 1, 2024 to June 30, 2025.

SWE leaders have engaged with members individually to encourage them to pursue leadership roles. While these one-on-one conversations play an important role in increasing diverse representation, there remains a need to enhance communication and marketing strategies to encourage more diverse candidates to volunteer for leadership positions.

In 2022, the board of directors established a special director position to offer recommendations for increasing diversity in leadership. Resulting recommendations included changing how information is communicated across the membership and adjusting the evaluation criteria for candidate selection. In response to these recommendations, the nomination committee now employs a diversity matrix to ensure a diverse slate of nominees for leadership positions. Another visible change that has taken place in response to recommendations has been in SWE's promotional materials. SWE members from all backgrounds are now prominently featured, including in SWE's publications, newsletters, and event marketing. This change allows SWE's membership, at all stages of their career journey, to see themselves as part of its community and leadership.

As SWE continued its DEI&B journey, feedback from the town halls, member surveys, and the affinity group leaders indicated there was a desire for SWE to show its commitment to DEI&B by having a dedicated staff member. Affinity groups in particular, especially those focused on DEI&B, were asking for more support from headquarters. SWE also sought to increase its DEI&B content for staff, members, and leaders. In addition, SWE's role in various DEI&B grant collaborations was growing.

In acknowledgment of the importance of diversity and inclusion to the organization, the board of directors approved the hiring of a DEI&B subject matter expert. In February 2022 SWE's first DEI&B manager was hired. This role serves as the primary point of contact for DEI&B-related efforts, including designing and developing content and programs, conducting gap analysis, analyzing program data and survey feedback, building and maintaining collaborative relationships with internal and external stakeholders, supporting DEI&B grants on which SWE is a partner, and serving as the liaison for DEI&B affinity groups. Ultimately, the DEI&B manager makes sure that everything that SWE does supports a welcoming and inclusive environment for members, partners, and other stakeholders.

Transparency and Accountability

While the marketing strategy that SWE initiated to promote its offerings was a critical component of its implementation efforts, transparency about the organization's progress toward achieving its DEI&B goals was vital in leadership accountability. Multiple focus groups held in 2020 to 2021 indicated that SWE's membership base was unaware of the DEI&B initiatives that SWE was putting in place. To address this, SWE's chief executive officer and the board of directors president began recording yearly videos with an update for members to hear directly from leadership about the work being done to increase diversity and inclusion within the society. However, SWE soon discovered that this communication method was not a very successful strategy, so in 2023, leadership published its first annual DEI&B report. The report focused on SWE's DEI&B efforts, with specific attention to its strategies to foster a culture of belonging.

The annual report provides a comprehensive overview of SWE's DEI&B initiatives. It details the organization's strategic goals and objectives, progress made over the past year, and areas requiring further attention. The report includes quantitative data from the KPI dashboard, offering metrics on member demographics, leadership diversity, and participation in SWE programs. Additionally, it features qualitative insights from focus groups and member surveys that highlight personal experiences and feedback. The report also outlines specific strategies and actions that SWE has taken to foster a culture of belonging, such as changes in nomination processes, targeted communication efforts, and collaborations with other diversity-serving organizations. SWE is committed to holding the organization accountable by publishing yearly DEI&B reports, each with a unique theme that captures progress and highlights areas for organizational growth in fostering a culture of belonging. With the annual report SWE aims to maintain transparency, encourage continuous improvement, and reinforce leadership's commitment to creating an inclusive and equitable environment for all members.

In addition to the annual report, SWE publishes ad hoc reports that provide insight into organizational culture. For example, the results of a member-focused culture study conducted in 2023 were shared with members in a report, blog post, and on social media to highlight areas of opportunity for improvement. The results reflected the importance of analyzing the experiences of members within specific subsets of the organization, including race/ethnicity, LGBTQ+, disability status, and career level to understand feelings of inclusion and belonging. Sharing these results broadly offers a measure of accountability for SWE leaders and provides increased understanding to address specific challenges.

In 2023, SWE hired a director of content strategy to revamp its communications strategy, including its member newsletters, website, social media, and publications. The director of content strategy assists leadership and staff in developing methods to enhance transparency regarding SWE's DEI&B efforts and to spotlight relevant data and metrics. Effectively communicating achievements and areas needing improvement is important to building trust in the process. These communication channels provide opportunities for SWE to share the progress toward its DEI&B goals and present an honest evaluation of where gaps exist as well as the society's plans to address them.

As part of SWE's educational journey in DEI&B, the organization has leaned into its partnerships with other diversity-serving organizations, including Advancing Indigenous People in STEM (AISES), the National Society of Black Engineers (NSBE), the Society of Hispanic Professional Engineers (SHPE), and the Women in Engineering ProActive Network (WEPAN). One notable collaboration involves a grant-funded project to support increasing the representation of women of color in the engineering workforce. The Women of Color in Engineering Collaborative (WCEC) is designed to improve the retention of diverse women in the engineering field, and entails sharing data and metrics, establishing goals, and monitoring progress across the collaborative. SWE's involvement in such projects underscores its commitment to addressing systemic issues that impact working engineers beyond its own membership, which is essential for advancing DEI&B within the engineering profession.

SWE holds Memorandums of Understanding (MOUs) with other diversity-serving organizations to solidify its partnerships and better support its members. These MOUs provide joint memberships for those interested in belonging to multiple societies, so members are not forced to choose which organization to devote their time and resources to; rather, SWE encourages its members to engage with other organizations in acknowledgment of the intersecting identities that its members hold. Through these collaborations, SWE has organized listening sessions with members from other organizations to better understand the challenges and needs of women in engineering, both within and outside its membership. This approach allows SWE leaders and staff to address questions, consider different perspectives, and incorporate feedback into SWE's activities. These listening sessions are another platform on which to share SWE's progress and initiatives toward its DEI&B goals.

SWE also hosts town hall meetings with its members to learn about their experiences, particularly regarding inclusion and belonging to the organization. SWE's board of directors has been strategic about what it does with this information, operating under the belief that those experiencing a lack of inclusion and belonging should not be the people assigned to fix the problems—in other words, "fix the system, not the women." Gathering feedback from members is an important part of the iterative nature of the DEI&B implementation process. Leaders must hear what issues are affecting members' experiences and consider solutions offered, but ultimately, it is up to the board of directors to drive the systemic changes needed across the organization that are causing these negative experiences. Leadership understands that addressing DEI&B goals is not a one-and-done effort.

Over time, as SWE has increased transparency around its DEI&B efforts, members have become more comfortable sharing their perspectives. With the changes implemented after the DEI self-assessment in 2021, member and leader attitudes and behaviors are slowly changing. While the board has been at the forefront of implementing these changes, SWE committees are now more seriously incorporating DEI&B into their operations and tying strategies to desired outcomes. Staff have modified their interactions with members and colleagues to foster a more inclusive and equitable work culture. For example, staff meetings routinely begin with a DEI&B moment to foster discussion and set the expectation of respectful dialogue and communication during the meeting. Leaders have also intensified their efforts to ensure that diverse representation is evident in more areas across the organization.

These expectations are extended to employer partners as well. SWE is working toward increasing DEI&B within the organization and wants to work with others who share their values. For example, SWE expects employers who nominate employees for SWE awards to present a diverse slate of candidates that reflect the society's global membership. SWE's employer partnership meetings always begin with a DEI&B moment, mirroring SWE's internal practices. Employer partners also receive SWE's reports on the society's DEI&B progress, as an example of the progress that can be made when organizations embrace DEI&B accountability and transparency.

Discussion and Reflection

Similar to other organizations that have spoken over the years about the importance of diversifying STEM professions, listed diversity as a priority, and highlighted DEI&B program offerings, SWE has espoused diversity as a core value for decades. But despite SWE's long history of supporting gender equity in engineering, we have only recently been able to track our journey toward creating a more diverse and inclusive environment within our organization. We are now in a better position to advocate for DEI&B because we put our words into action and hold ourselves accountable. Accountability requires consistency, transparency, grace, and patience in acknowledging that this is a work in progress.

SWE has taken significant steps to broaden the scope of data in its DEI&B analysis. Through strategic partnerships and data standardization, the organization has worked to streamline the collection of key demographic information across all programmatic areas. This inclusive approach allows SWE to comprehensively understand its membership while considering unique experiences and needs within its diverse community. Additionally, SWE has introduced KPIs to evaluate progress and impact through various lenses to ensure a more nuanced assessment of our efforts. SWE's dedication to expanding the data scope equips leadership to make informed decisions, customize initiatives, and promote a more equitable and inclusive engineering industry for everyone.

SWE has sought to foster a more comprehensive and nuanced understanding of our membership. If our members feel that they are not welcome within the organization, then our initiatives to create a more diverse and inclusive society lose their significance. The quantitative and qualitative data collected and reported informs the organization's decision-making processes, program development, and initiatives, which allows us to create more inclusive and tailored experiences for our diverse community by helping us understand what our members are really experiencing and responding to that feedback. While much of this work is happening internally, sharing these activities and the progress made toward DEI&B goals with members and the public helps ensure accountability through transparency. We now share what is working and what is not through our newsletters, publications, annual DEI&B report, and other avenues.

However, the journey has not been without challenges. Listening to members' experiences can be difficult for both the speaker and listener but providing opportunities to be open and honest is necessary to identify where problems exist so they can be addressed. Figuring out how to reach members, gather information and feedback, and share progress toward DEI&B goals has been a continuous learning process. Our membership is growing to include individuals from different geographies and generations, and how we obtain information and engage with one another is constantly shifting. Overcoming the barriers of "this is the way we have always done it" is challenging. Without leadership support, the integration of DEI&B into SWE's strategic plan, and the assignment of responsibilities to specific individuals and committees, it would have been easy to make only minimal efforts toward addressing the issues that hinder our goal of becoming a truly diverse and inclusive organization.

In alignment with the iterative process required to achieve culture change, we continue to analyze the data and make necessary changes to our strategies. As noted earlier, adjustments to the FY 2024 annual member survey in measuring belonging have revealed that some members do not feel as strong a sense of belonging at the local level as they do at the society level. To address this, SWE has created metrics for FY 2025 to evaluate local group health and has developed strategies to increase belonging at the local level. One such strategy is the recent incorporation of section relations into SWE's staff responsibilities to provide direct support to local sections.

Another example of the changes in response to feedback from members through our data measures can be seen in our transparency efforts. For example, many members remain unaware of the changes to SWE's awards and leadership pipeline processes. SWE FY 2025 plans include improved marketing strategies and more frequent metrics reporting to members and staff to keep all stakeholders informed of the progress being made toward our DEI&B goals.

This case study offers insights into SWE's experiences with increasing DEI&B within the organization over the past ten years. Professional societies interested in taking a more active role in moving from a passive to an active promoter of DEI&B within their organizations should consider:

- What exactly do we want to achieve? Identify the goals that you want to reach.
- Do we have leadership support to allocate resources for DEI&B strategies and make changes if certain strategies are not working as planned?
- Are all the necessary stakeholders included in our plan, from leadership to membership to staff? For example, are there ways to gather different perspectives by procuring honest feedback from various stakeholders?
- Is my organization prepared to adjust its data collection strategies, including how, when, and what data is collected?
- How will my organization share what we are doing with our stakeholders, including the goals we have set and the progress we are making toward those goals?

Concluding Remarks

As SWE prepares to celebrate its seventy-fifth anniversary in 2025, we reflect on our challenges and successes since our founding. As a diversity-serving organization focused on gender equity in STEM, the society has experienced its share of pushback. Throughout it all, we have remained focused on our mission to help women in engineering and technology succeed and advance in their personal and professional endeavors. We welcome everyone, including allies and male champions, as we know gender equity is unattainable without such support. Ultimately, SWE's purpose is to support women—all women—and that is what we aim to do with every program, service, and support we provide.

Promoting accountability and transparency within SWE puts the organization on a path to ensuring DEI&B efforts are not merely superficial gestures but are deeply ingrained in the society's values and practices. Setting clear goals, defining measurable objectives, transparently communicating both progress and challenges, and sharing expectations with SWE partners all contribute to building greater trust within the larger community. SWE has gone beyond telling others what they should be doing to increase DEI&B in the engineering profession; SWE is leading by example.

We recognize that SWE is a safe space for many women in engineering and technology. We want SWE to be a constant that they can rely on, regardless of what happens in other spaces. To this end, SWE is committed to leveraging data to drive meaningful changes to support our members better, advance equity and inclusion, and ensure that the society remains at the forefront of promoting diversity in engineering and technology fields.

Note

1 The authors thank Troy Eller English, SWE's archivist, for her meticulous documentation of the society's historical records that provided the background for this chapter. We thank Karen Horting—SWE's chief executive officer and executive director—for her support and assistance while writing the case study. We thank Sohn Cook for reviewing the chapter for grammar and clarity as well as Honna George—who helped drive much of the DEI&B efforts at the organizational level—for her review and support.

References

Bezrukova, K., C. S. Spell, J. L. Perry, and K. A. Jehn. 2016. "A Meta-analytical Integration of over 40 Years of Research on Diversity Training Evaluation." *Psychological Bulletin* 142 (11): 1227–74. https://doi.org/10.1037/bul0000067.

Campbell-Montalvo, R., A. L. Putwen, L. Hill, H. E. Metcalf, E. L. Sims, J. W. Peters, A. N. Zimmerman, D. L. Gillian-Daniel, G. M. Leibnitz, and V. A. Segarra. 2022b. "Scientific Societies Integrating Gender and Ethnoracial Diversity Efforts: A First Meeting Report from Amplifying the Alliance to Catalyze Change for Equity in STEM Success (ACCESS+)." *Journal of Microbiology and Biology Education* 23(1): e00340–21. https://doi.org/10.1128/jmbe.00340-21.

Chavez, C. I., and J. Y. Weisinger. 2008. "Beyond Diversity Training: A Social Infusion for Cultural Inclusion." *Human Resources Management* 47 (2): 331–50.

Corbett, C., and C. Hill. 2015. *Solving the Equation: The Variables for Women's Success in Engineering and Computing*. American Association of University Women.

Doty, H. 2021. "Updating SWE's Diversity, Equity, and Inclusion Statement." *SWE All Together*, July 2. https://alltogether.swe.org/2021/07/updating-swes-diversity-equity-and-inclusion-statement/.

Eller English, T. 2023. "What's Past Is Prologue: The Future Is Ours to Write." *SWE Magazine*, May 10. https://swe.org/magazine/past-is-prologue/.

George, H., N. Mohan, and R. Rincon 2023. *Fostering Belonging: Diversity, Equity, Inclusion & Belonging Organizational Initiatives at SWE*. Society of Women Engineers. https://swe.org/download/33774/?tmstv=1724252014.

Gienapp, A., and C. Hostetter. 2022. *Developing a Theory of Change: Practical Guidance*. Annie E. Casey Foundation. https://www.aecf.org/resources/theory-of-change.

Madzima, T. F., and G. C. MacIntosh. 2021. "Equity, Diversity, and Inclusion Efforts in Professional Societies: Intention versus Reaction." *The Plant Cell* 33 (10): 3189–93. https://doi.org/10.1093/plcell/koab186.

Meyer-Gutbrod, E., L. White, B. Schieler, M. Behl, and S. Park. 2023. "The Role of Professional Societies in Advancing Diversity, Equity, Inclusion, Justice, and Accessibility in the Fields of Coastal and Ocean Science." *Oceanography* 4 (36): 124–31.

Mohan, N., R. Rincon, and T. Swanigan. 2024. *Embedding DEI&B in All We Do. The 2024 SWE Report on Diversity, Equity, Inclusion, and Belonging*. Society of Women Engineers. https://swe.org/wp-content/uploads/2024/05/24-SWE-GL-021_DEIB_2024Report_101524_jc9.pdf.

NASEM (National Academies of Sciences, Engineering, and Medicine). 2021. *Transforming Trajectories for Women of Color in Tech*. The National Academies Press.

Shiffman, D. S., T. A. Álvarez, C. W. Bangley, R. Boyt, I. M. Côté, T. S. Daly-Engel, A. C. D. Davis, L. C. Gaskins, J. Graham, R. T. Graham, S. Johri, C. C. Macdonald, E. W. M. Paig-Tran, A. I. Roca, G. D. Schwieterman, L. B. Whitenack,

T. R. Wiley and L. A. Ferry. 2022. "What Can Professional Scientific Societies Do to Improve Diversity, Equity, and Inclusion: A Case Study of the American Elasmobranch Society." *Frontiers in Education* 7: 842618. https://dx.doi.org/10.3389/feduc.2022.842618.

SWE (Society of Women Engineers). 1952. *Certificate of Incorporation 35610*, February 13. https://swe.org/wp-content/uploads/2019/06/swe_articles_of_incorporation.pdf.

SWE (Society of Women Engineers), host. 2020. *Diverse: A SWE Podcast*, podcast, "SWE Stories: Tales from the Archives: SWE Celebrates 70th Anniversary," May 27. https://soundcloud.com/swepodcasts/swe-stories-tales-from-the-archives-early-achievement-award-recipients-offer-advice-pt-3.

Taplin, D. H., and H. Clark. 2012. *Theory of Change Basics: A Primer on Theory of Change*. ActKnowledge. https://www.theoryofchange.org/wp-content/uploads/toco_library/pdf/ToCBasics.pdf.

9

Investments toward Justice, Equity, Diversity, and Inclusion over a Decade of Association for Advancing Participatory Sciences Conferences

Jacklyn S. Kelty, Abigail Hils, Linda Silka, and Alison Cawood

Conferences provide critical opportunities for members of scientific associations to network, share discoveries, give peer feedback, and pursue life-long learning (e.g., Edelheim et al. 2017; Etzion et al. 2022).[1] Gathering members and guests with shared interests can be rewarding for the attendees. On the other hand, the tight alignment of people and content can form echo chambers by amplifying ideas that already circulate within the association. Conference organizers can promote the exchange of ideas with the broader community by actively welcoming connected disciplines and institutions. The inclusion of outside perspectives also widens the scope of knowledge shared at conferences. For scientific communities to thrive, associations must pursue inclusive conference planning techniques that can support diverse perspectives and diverse attendees. Unfortunately, conference organizers often overlook their attendees' expertise and passion for movements toward equity and justice. For scientific communities to maintain relevance, conferences must authentically produce equitable and just outcomes that validate the substantial investments made by attendees, members, hosts, conference organizers, and sponsors. Scientific associations cannot afford to neglect the fundamental values of justice, equity, diversity, and inclusion during conference planning.

The Association for Advancing Participatory Sciences (AAPS) was conceived by hosting a conference in 2012. The association communicates developments in the field by hosting regular in-person, virtual, and regional conferences. In deference to our members' decisions (Kelty et al. 2022; Putnam 2023), we will use "AAPS" to refer to the association (formerly the Citizen Science Association). AAPS has set "advancing knowledge through research and monitoring done by, for, and with members of the public" as its core purpose (AAPS 2024). Membership and activities are not limited to specific scientific disciplines. Members' fields of expertise range across natural sciences, engineering, education, social sciences, humanities, and cross-disciplinary research. AAPS brings together avocational researchers, academics, regulatory agencies, nonprofit organizations, and community organizers. New directions for AAPS develop by interweaving contributions from members whose missions are as divergent as Indigenous knowledge keeping and online crowdsourcing innovation. AAPS provides a rare forum for its members to advocate for justice, equity, diversity, and inclusion, a forum that discipline-specific associations often lack.

This chapter assesses AAPS prioritization of the fundamental values of justice, equity, diversity, and inclusion during the conferences between 2017 and 2022. These conferences were high-priority opportunities to magnify and distribute emerging justice, equity, diversity, and inclusion innovations. This chapter explores how and why AAPS prioritized specific actions in response to coinciding circumstances, its available resources, the risks and benefits of these investments, and the lessons learned. The nimbleness of the AAPS conference organizers and frank communications from AAPS members inspired substantial transformations over the years. The authors synthesized this case study narrative through a literature review of public and internal conference documents with enrichment from their experiences as conference organizers and facilitators. This manuscript gives the AAPS perspective on the question: "Does the achievement of justice, equity, diversity, and inclusion require incremental adaptation to shifting contexts by conference organizers or is the reinforcement of previously implemented strategies enough?"

Background for the Study

The AAPS chartered the Integrity, Diversity, and Equity Working Group in 2015 to assess association practices and infuse new ideas into the organization. The working group went into a hiatus after the 2017 conference and was reassembled after the 2019 conference. In 2020 the operating model was changed to distribute justice, equity, diversity, and inclusion responsibilities among all working groups. The transition was a response to calls-to-action produced by the deliberations of various AAPS working groups related to the 2020 Black Lives Matter movements and Covid-19 pandemic injustices. Additionally, the name was changed to the Justice, Equity, Diversity, and Inclusion Working Group to distinguish its function from the ethics and environmental justice teams. Currently, each working group is expected to invest in tangible actions to build toward the AAPS fundamental values. Overall, this shift can be understood as moving from a watchdog working group to an organizational approach that holds each member and leader accountable to the fundamental values of AAPS.

Harmonized definitions were developed in 2022 for the AAPS fundamental values, process principles, and research relationship outcomes (Kelty et al. 2022). AAPS swiftly directed working groups to implement these concepts within their own contexts. The definitions emerged through a crowdsourcing process that involved over fifty volunteers that responded to an AAPS call-to-action, the AAPS Justice, Equity, Diversity, and Inclusion Working Group, and the AAPS Environmental Justice Practitioners Working Group. These definitions are informed by, but do not fully address, cultural- or identity-based lenses such as Indigenous, antiracist, or feminist views. The Justice, Equity, Diversity, and Inclusion Working Group assembles every other year to maintain the AAPS harmonized definitions. The following fundamental value definitions represent the current understanding within the field represented by AAPS.

- *C*science*—This represents the many research practices that are welcome in AAPS, including but not limited to community research, citizen science, hood research, co-created science, gamification, communities of practice, contributory science, collaborative science, patient-centered outcomes research, community-based action research, democratized science, collective and distributed intelligence, citizen sensing, service learning, makers and hackers, community-engaged research, volunteered geographic information, participatory research, open science, user-centered design, amateur science, radical science, grassroots activities, public participation in scientific research, democratic governance of science, crowdsourcing, and community-owned and -managed research.
- *Justice (J)*—The fair distribution of rights, power, services, opportunities, and risks between individuals and communities, including restoration from past systemic harms.
- *Equity (E)*—The promotion of just institutional processes that recognize historical and enduring exclusionary practices and acknowledge each individual's circumstances by allocating the resources and opportunities needed to reach an equal outcome.

- *Diversity (D)*—The mosaic of personal attributes, social affiliations, ethnic backgrounds, and ideological orientations within a group of people.
- *Inclusion (I)*—The intentional creation of power structures in which individuals feel safe, welcome, valued, respected, and supported *as their authentic selves* in their participation and decision-making.
- *Funding ($)*—The resource and financial investments that are necessary and critical to abolish persistent academic research exploitation, the misplacement of public trust in exclusionary institutions, and the enduring harms of unjust systems.

The authors contributed substantial volunteer efforts to build AAPS. *Kelty* is a white woman from an economically and educationally privileged family in Los Angeles. Her academic, community-based, participatory action research seeks environmental justice for fence-line communities and workers exposed to industrial pollution. She served as co-chair of the AAPS Justice, Equity, Diversity, and Inclusion Working Group from 2020 to 2024. *Hils,* of mixed heritage and from a family with economic and educational privileges on Turtle Island, holds an academic background in marine science and science communication. Their career was principally dedicated to advancing climate justice for coastal communities and building the knowledge systems essential for the revolutionary change envisioned by the United Nations. In 2021 they served on a conference working group and took on a leadership role in planning the 2022 conference. *Silka* is a white woman whose father was the first in his family to have the opportunity to go to college through the GI Bill (Servicemen's Readjustment Act of 1944, Pub. L. No. 78-346, 58 Stat. 284). As a community psychology college professor, Silka facilitates community–university research partnerships, often with refugee and immigrant communities. She has served on every AAPS conference working group from 2012 to 2022. *Cawood* is a white woman from an economically and educationally privileged family in South Carolina. She has an academic background in marine science and spent most of her career working at the interface between communities and traditional research. She also served on every conference working group from 2015 to 2019, led the conference planning teams from 2021 to 2023, and currently serves on the AAPS board of directors. The interdisciplinary values of AAPS are exemplified by the diversity in our research disciplines, occupational duties, and career stages. The authors are committed to building capacity and opportunities for AAPS researchers and participants.

Case Study Narrative

This section outlines the 2012 and 2015 conferences that set the stage for subsequent justice, equity, diversity, and inclusion developments from 2017 to 2022. The origins of AAPS trace back to the 2012 Conference on Public Participation in Scientific Research held as a supplement to the ninety-seventh annual meeting of the Ecological Society of America (Benz et al. 2013; Heimlich 2012). The gathering aimed to: (1) formalize c*science as a field of study and practice; (2) stimulate communication, collaboration, and innovation; and (3) develop an organizational structure for the field of c*science, including a professional association, regional networks, a journal, and biannual meetings (Benz et al. 2013). However, the 284 conference attendees were notably homogeneous: 80 percent were women, 72 percent were adults, and 88 percent were white (Heimlich 2012). From conception, AAPS recognized this lack of diversity as a threat to the relevance and potency of the new scientific association.

Leaders who emerged from the 2012 conference went on to form AAPS. The first official AAPS conference was held in 2015 as a pre-conference to the American Association for the Advancement of Science annual meeting. The goals for the conference, as identified by the steering committee chair, were to "enhance communication, collaboration, and innovation in this field of extreme interdisciplinarity" (AAPS 2015a). By then, AAPS had grown to over 3,000 members and launched a peer-reviewed journal (AAPS 2015a). One of the six themes for the 2015 conference was "broadening engagement to foster diversity and inclusion" and was defined by conference organizers:

> As a broad field, inclusive of a variety of philosophies and approaches to participatory science, [c*science] comprises a diversity of: (1) disciplines (e.g., public health, ecology, astronomy, social justice); (2) research traditions (e.g., traditional ecological knowledge, traditional citizen science, water-quality monitoring, participatory rural research); (3) project models/approaches (e.g., contributory, collaborative, co-created); and (4) roles (e.g., policymakers, project coordinators, researchers, community volunteers). Yet research shows that many groups are under-represented in these efforts. How can [c*science] increase inclusion and foster a greater diversity of participants, so that access to and benefits of projects can be spread more broadly?
>
> (AAPS 2015b)

Three panels, two presentation sessions, and four symposia were organized under this theme to fill about 18 percent of the total presentation time at the conference (AAPS 2015a).

Despite these efforts, the 2015 conference evaluation revealed ongoing challenges. While the conference attracted over 600 attendees from twenty-five countries, the diversity of attendees remained limited. Free-response comments in the evaluation report discussed the lack of diversity among attendees, with particular reference to the lack of representation in a session focused on c*science among Indigenous communities (Heimlich and Timko 2015). Attendees included c*science participants, project leaders, educators, scientists, technology specialists, evaluators, and others. A wide range of research disciplines were represented including (but not limited to) astronomy, molecular biology, human and environmental health, psychology, linguistics, environmental justice, biodiversity, conservation biology, public health, genetics, engineering, cyber technology, and human–computer interactions (AAPS 2015c). The evaluation report for the 2015 conference lacked any measurement of community-led c*science perspectives; it is unlikely that these practitioners were well represented at the conference since nearly 40 percent of attendees were from government or academic institutions.

Conference evaluations from 2012 and 2015 did not explicitly address the exclusionary practices of the academic conference spaces, which were comfortable for the initial founders but less so for those outside established professional networks. As the field continued to grapple with the power dynamics inherent in contributory versus collaborative c*science projects (Eitzel et al. 2017; Shirk et al. 2012), the need for a more inclusive approach became increasingly apparent. In response, conference organizers started conducting a needs assessment with AAPS membership before each conference, aiming to align the events with the community's evolving expectations. Pre-conference assessments identified a desire for conferences that prioritized skill-building, featured diverse perspectives, emphasized discussion rather than a unidirectional flow of information, and the specific inclusion of members of related fields such as educators and environmental justice practitioners (Hils et al. 2023; Duncan et al. 2024). The volunteer conference working group collaborated with local hosts to address these needs as AAPS endeavored to create a more inclusive and representative environment for all participants.

The early AAPS conference planning and implementation attempts reflect the broader commitment within the c*science field to integrate justice, diversity, equity, and inclusion principles into core community practices. The lessons learned from the early conferences laid the groundwork for more inclusive and equitable gathering in subsequent years as summarized in this case study for 2017 to 2022. This case study is one chapter of the ongoing efforts to make c*science into a field that truly reflects the diversity of its global community.

Findings

Justice

The 2019 conference was a pivotal spotlight for justice in AAPS. The conference intentionally highlighted environmental justice practitioners as thought leaders and critical audiences for c*science (AAPS 2019). The focus on environmental justice was supported by the local host's existing ties to environmental justice practitioners. North Carolina was the birthplace of the American environmental justice movement that extended the activism of the Civil Rights Movement into the next generation of leaders (Geiser and Waneck 1983). In 1982 Warren County leaders successfully opposed the government's use of racist environmental and community development research to justify the location of a hazardous waste facility. Environmental justice practitioners are just as powerful in North Carolina today. For example, the West End Revitalization Association (WERA) leverages c*science to improve their community through regulatory agency channels (Heaney et al. 2007). WERA is often sought by these agencies to serve on advisory boards and to provide justice, equity, diversity, inclusion, integrity, and funding training to agency staff. The bountiful history of environmental justice in North Carolina strongly influenced the choice of Raleigh as the venue for the 2019 conference.

AAPS leadership and conference organizers acknowledged that tangible support would be required to facilitate conference attendance for environmental justice practitioners. AAPS was not a proven ally to the environmental justice community, so conference organizers needed to reveal clear conference attendance benefits that could offset the distraction from practitioners' existing activities. AAPS board members Caren Cooper and Sacoby Wilson were awarded a $74,800 National Science Foundation travel grant (NSF 1842188) to support conference participation for seventy-five environmental justice practitioners. The attendees that took advantage of this opportunity were grateful for the reduction in financial barriers, but concerns emerged about the short turnaround for applications and shortfalls related to accommodations. For example, travel award recipients were expected to share a hotel room, which made some attendees uncomfortable and limited their ability to recharge in this already unfamiliar environment. AAPS took this feedback seriously and transitioned to a different approach to subsidize conference attendance for the next in-person conference in 2023. AAPS created scholarships using contributions from donors, members, and sponsors to provide funds for people that could not otherwise afford to attend. The AAPS Environmental Justice Practitioners Working Group (founded through the 2019 conference) helped to spread the word about scholarship availability. The 2023 scholarships allowed participants to make their own travel arrangements and provide an up-front cost estimate to the scholarship committee. Depending on the needs of individuals, funds were either provided up-front or when the person got to the 2023 conference. Scholarship recipients were no longer required to submit receipts or additional paperwork in order to access funds.

The AAPS Code of Conduct was developed by the conference working group in 2019 and introduced throughout AAPS activities (AAPS 2021a). The simultaneous efforts to actively invite more diverse practitioners to the conference and launch the code of conduct caused unintended tensions. AAPS Code of Conduct developers did not foresee all the complicated layers of interpretation and implementation that should be carefully considered before launch. Some environmental justice practitioners and other attendees from historically excluded groups felt targeted for censorship in the conference space that was formerly dominated by people that identify as academics, white, and women. Specifically, suspicions emerged that the code of conduct was targeted toward attendees from historically excluded groups because AAPS leadership and conference organizers assumed these attendees would be unaware of "correct" ways to engage in an exclusive academic space. An even wider group was concerned that attendees would be banned from the conference, reprimanded, or otherwise punished for any criticism of the status quo. Many found the code of conduct unclear with regard to who was enforcing the code, the processes for response to reports, what the AAPS-specific consequences would be for violation, and what professional and personal implications could result from violations or reporting violations. AAPS learned from this experience and updated the code of conduct for future conferences.

An environmental justice strand ran through the 2019 conference sessions. An environmental justice panel was prominently featured as part of the conference. This event required separate tickets to attend in person, but the recording was made available publicly (Wilson 2019). For some, this was the highlight of the conference. Overall, evaluation results indicated that the strand was well received and perceived as inclusive. However, there were some—particularly from those more familiar with academic conferences—who felt this environmental justice content to be divisive, believed the AAPS conference was not supposed to be a social science conference, and desired to focus content on the technical science conducted in communities and less about translational science or the communities themselves. These conference evaluation respondents felt that the environmental justice focus detracted from technical science. Some technical science presenters felt sabotaged by unexpected audience questions and criticisms that dug deeper into the ethics and implications for their research instead of staying narrowly focused on the specific content presented.

Conversely, some of the environmental justice practitioners reported that the AAPS conference did not measure up to events designed for environmental justice communities because of the more academic focus. Root causes identified by the environmental justice practitioners related to the minimal input they were invited to provide throughout the planning process, the focus of content on problems rather than solutions, the limited scope of communities represented, and a sense that the environmental justice practitioners and community members were not well integrated with the more academic and traditional conference goers. Environmental justice practitioners organized to form an AAPS working group to advocate for their role in this organization and conferences moving forward. This movement was wholeheartedly supported by AAPS leadership. As of the 2024 submission of this chapter, the AAPS director and board members continue to actively participate in the Environmental Justice Practitioners Working Group meetings every other week.

AAPS organizes conference field trips to steer the expertise of attendees toward justice issues that are relevant to the host communities. In 2019 one of the five field trips offered to conference attendees was hosted by WERA to highlight the utility of c*science to support local environmental justice activism. The other four field trips offered in 2019 were nature adventures to local wilderness sites and the natural science museum. This pattern was similar to the 2017 conference when one of three field trips was focused on justice in the Twin Cities. Conference organizers subsequently refined the field trips to yield authentic benefits for the communities that host spectators and to prevent extractive approaches to toxic tours (Chew Bigby et al. 2022). In 2023, community-embedded activities provided opportunities for conference attendees to genuinely connect with local libraries and community-based organizations seeking subject matter experts or opening their doors to share their c*science experiences. AAPS conferences continue to move these valuable experiences toward idea-to-action and workshop formats that facilitate just outcomes instead of the extractive approaches that historically drain knowledge and resources from host communities. Efforts to fairly distribute power, services, opportunities, and risks among presenters and attendees evolved over the years of AAPS conferences to improve practices and heal from harms caused by previous conference or discipline-specific injustices.

Equity

The national organization invested in and nurtured members within shared interests and spaces by hosting a virtual "Connected Where You Are" conference in 2022 with anchored regional events. Organizers embraced a decentralized yet cohesive structure for this event to address isolation and disconnection that emerged from the Covid-19 pandemic. This hybrid model facilitated meaningful engagement for a diverse range of participants by overcoming geographical and financial barriers. Local hosts were able to link their events to the AAPS conference to receive logistic assistance and promotion. For example, the University of California, Davis Center for Community and Citizen Science and the California State University, Bakersfield Citizen Science Project hosted a three-minute pitch event for anyone that wanted to share, regardless of AAPS conference attendance or membership.

In 2022, membership and conference fee scales were introduced based on an individual's role within c*science and the size of their organization. This approach to fees enhanced access and participation opportunities beyond the previous reduced fee categories for formally enrolled students, early registrants, and attendees from lower-income countries. These changes demonstrate a commitment to accessible and inclusive conference environments where financial constraints would not impede individuals from contributing to, and benefiting from, the conference experience. However, a few attendees exploited these systems to pay nothing for attending the conference despite having the means to contribute. For example, AAPS did not require substantial documentation to utilize these financial accommodations so some individuals with grant funding that belonged to powerful institutions were able to use these accommodations instead of institutional or grant support. AAPS conference organizers and leadership agreed that the benefits to accessibility overwhelmed the negative impacts of the few attendees that may have intentionally taken advantage of the financial accommodations (see the "Justice" section for additional context about stipends and conference fee waivers). As the 2025 conference registration is open at the time of this chapter submission, reduced rate tickets are available for the following self-reported attendees: (a) students, (b) K–12 formal classroom teachers and educators in afterschool programs, (c) retirees, (d) environmental justice practitioners and Indigenous practitioners, (e) project volunteers, and (f) registrants who are based in countries classified as the Global South by the United Nations Educational, Scientific, and Cultural Organization. Additionally, discounts were made available for volunteering for five or more hours to direct programmatic support. Critically, conference organizers reduced the administrative burden to request support by committing to a single application for any or all conference fee–discount options.

The most significant challenge of the 2022 conference was the organizer burnout, exemplified by the drastic reduction in the poster committee from nine members in 2021 to two in 2022. Hosting such a large, continent-wide, international event proved logistically complex and demanding in terms of volunteer time, energy, and resources. This was further compounded by the need to continuously adapt to unforeseen challenges while also creating a more inclusive space for nominations in multiple languages. This combination of factors led to significant fatigue among the team. In hindsight, this experience highlights the importance of planning for the well-being of the organizing team just as carefully as the event itself. Rethinking traditional models may be required to mitigate the exhaustion that so often accompanies planning large, international conferences.

Equity and inclusion principles were central to the 2022 agenda in response to feedback from AAPS members and previous conference attendees. The three tracks for the conference were "Future of C*Science," "Working Across Boundaries," and "Equity and Inclusion in C*Science." The organizers prominently posted the following equity statement on the conference website and shared this context in conference forums:

> [AAPS] is committed to providing an equitable and inclusive virtual conference and celebrating the diverse voices in our space. [AAPS] strives to ensure that all our members and participants feel welcome and comfortable sharing their insights and contributions with the larger [c*science] community, and that all our participants can meaningfully engage in our events.
>
> (AAPS 2022)

Each presenter was required to specifically disclose equity, diversity, and inclusion-related considerations by responding to an abstract submission prompt. Conference organizers provided support for presenters who were unfamiliar with how to effectively respond to this new requirement. Poster peer review was available before the event to assist with incorporation of these disclosures into the final presentation. Speakers, largely selected based on nominations from AAPS members, represented a wide diversity of specialties, career stages, and roles within c*science. This nomination approach ensured that diverse voices and perspectives were not only represented but also actively heard and valued. After concerns were reported about the accessibility and intention of funding agencies that attended previous conferences, a critical spotlight was put toward equity for various governmental, nonprofit, and philanthropic groups in an "Innovations in Funding for C*Sciences" symposium. Further, the agenda was modified to feature nearly equal time each day for symposia, plenary speakers, facilitated poster sessions (Duncan et al. 2024), and ideas-to-action sessions. The ideas-to-action session format allowed participants to choose engaged and collaborative sessions for building skills, advancing efforts, seeking solutions, and other ways to leverage inspiration for impact. Equity was intricately woven into every aspect of the 2022 conference agenda structure, from abstract submissions to creating post-conference action plans.

The 2022 conference even served as a platform for AAPS to change its name due to member concerns about the exclusionary interpretation of "citizen science" in American communities. The complexity of terminology is a constant struggle that impacts many participants and practitioners in our field of extreme interdisciplinarity (Cooper et al. 2021; Eitzel et al. 2017; Ellwood et al. 2023). AAPS leadership unveiled a call-to-action to open formal discussions about an organization name change in response to previous working group conversations, conference presentations, and member feedback forums. "A Landscape of Terms for an Inclusive C*SA" plenary collaborative session was hosted by the AAPS director and other organizational leaders in our field. This 2022 conference discussion sparked leadership mobilization (Shirk 2022a, 2022b). For example, AAPS released a blog post to provide a summary of the responses from membership:

> Renaming an association should not be confused with either a new name for the field or a mandate for others to make change, nor should it be seen as an abandonment of the term that built this movement. As some have recently

> written, "citizen science" in global communities is recognized [as] an inclusive term for these practices. Others help us see that, for many (particularly but not exclusively in the [United States]), the term "citizen science" limits engagement and causes harm. These two things which may seem to contradict each other can both be true – the challenge is in how to operate in light of that complexity.
>
> (Putnam 2023)

AAPS leadership formalized the process by providing boundaries and requirements for a new name, then invited the full membership to propose names that were screened by the board, and finally distributed a formal membership survey in 2023. The membership decision to change the name was announced at the 2023 conference and the name change was completed by 2024 (Shirk 2024). The journey of the AAPS organizational name demonstrates how critical this high-visibility conference forum is for equity conversations across a whole field of study. The 2022 conference promoted just institutional processes that recognized historical and enduring exclusionary practices and acknowledged attendees' circumstances by allocating the resources and opportunities needed to reach an equal outcome.

Diversity

In early 2020, the onset of the Covid-19 pandemic brought unprecedented challenges for scientific conference organizers. AAPS was set to bring together diverse attendees at an in-person conference to be held at Arizona State University in 2021. As the severity and global scale of the pandemic was revealed, the AAPS faced a difficult decision: cancel the 2021 conference or adapt to the new reality. AAPS leaned on its innovative and resilient members to transform the conference into CitSciVirtual 2021, a pioneering virtual event that would go on to set a new frontier for future AAPS gatherings.

Conference organizers wanted CitSciVirtual 2021 to be more than just a contingency plan. Organizers embraced this moment as an opportunity to reimagine what a conference could be without physical gathering. In the field of c*sciences, there is great debate about how to maintain the highest level of active and productive participation that often touches on the benefits and weaknesses of in-person, hybrid, and virtual strategies. This subject matter expertise of the AAPS conference organizing team and membership allowed the organization to thoughtfully make this transition in ways that other discipline-specific organizations could not. Additionally, previous conversations about transitioning to a virtual format meant that a groundwork was already in place for a change from in-person to virtual (see the "Inclusion" section).

Themed "Local, Global, Connected," the event aimed to enhance accessibility and inclusivity by eliminating travel barriers, thus opening the doors to a wider audience. Participants from around the world—high school students, scientists, practitioners, activists, funders, policymakers, non-governmental organizations, and artists—could now join the conference without leaving their homes. The decision to accept posters in multiple languages and to host the conference over a month-long period underscored the event's commitment to broadening participation and allowing flexible engagement.

As the first fully virtual conference organized by AAPS, CitSciVirtual 2021 was a significant departure from traditional formats. Despite the challenges inherent in such a transformation, the event achieved many of its core goals (Hils et al. 2023). Over 700 participants from thirty-six countries came together, contributing to 240 digital posters as well as 55 workshops, keynotes, plenary discussions, and social events. The virtual platform facilitated valuable networking opportunities, skill-building sessions, and feedback mechanisms. Attendees appreciated the chance to connect with members of the c*science world, engage with a broad spectrum of projects, and participate in interactive, discussion-based sessions. Particularly impactful were the sessions focused on historically excluded groups, such as a panel featuring representatives from historically Black colleges and universities and a workshop that integrated Indigenous ways of knowing into c*science-project participation and development (AAPS 2021b).

The month-long duration of the conference, while intended to provide flexibility, proved overwhelming for some participants and led to a significant drop in attendance in the later weeks. Additionally, the abundance of sessions at different times reduced the urgency for attendees to participate in specific events, which contributed to lower engagement levels. Workshops, particularly those held later in the conference, often saw low attendance, and pre-recorded materials designed for asynchronous learning frequently failed to translate into meaningful interactions through discussion board and chat features. Feedback from attendees highlighted the need for more active and interactive elements as well as targeted support for emerging researchers who could benefit from structured networking and skill-building opportunities.

The virtual format of CitSciVirtual 2021 allowed for unique feedback mechanisms, such as asynchronous comments on digital posters, which facilitated global and multilingual engagement. However, broader feedback opportunities for other content types were less successful. Some participants did not fully engage with posters or collaborative sessions. Concerns about the visibility and permanence of feedback may have deterred open sharing. Furthermore, accessibility issues persisted, particularly with the CSA Connect platform, where many attendees struggled with navigation and engagement. For example, the platform was not compatible with screen-readers that are critical tools for attendees with visual impairments. These challenges underscored the need for more intuitive virtual platforms and better support systems to ensure a more inclusive and accessible experience.

Despite these hurdles, CitSciVirtual 2021 succeeded in expanding the conference's reach by drawing participants from diverse backgrounds, including historically excluded attendees such as high school students and artists. The event's inclusiveness enabled international engagement and provided opportunities for those with significant work or personal commitments. To accommodate different time zones, the conference featured pre-recorded content and live sessions at several times to deliver global accessibility.

CitSciVirtual 2021 was a groundbreaking event that brought together a global community and emphasized interaction and dialogue through discussion-based poster sessions and hands-on workshops. Efforts to ensure diversity and inclusion saw some success, but the virtual format did not always translate into meaningful experiences for attendees. Nonetheless, the conference set a new precedent for future events by emphasizing the synergy of virtual and in-person formats, with a commitment to continuous improvement in addressing duration, engagement, and platform issues for future gatherings. The Covid-19 pandemic-related restructuring embraced the strengths of virtual community building to help diversify the pool of potential attendees and to reduce barriers to attendance.

Inclusion

During the 2010 decade, several American states passed exclusionary laws criminalizing the use of bathrooms that did not align with the sex noted on people's government documents. These bans underscored the injustice of allowing some individuals to attend and benefit from conferences safely while others faced exclusion based on their gender identity. In response, other states and organizations enacted travel bans or otherwise boycotted these discriminatory laws to ensure the safety of conference attendees. Conferences represent a significant economic opportunity for host cities since scientific associations make substantial investments in conference logistics.

AAPS planned to hold the 2017 conference in North Carolina, a state with a strong history of participatory research where many members reside. However, North Carolina passed the Public Facilities Privacy and Security Act (HB2, 2016) that restricted bathroom use for transgender individuals about one year prior. Beyond the personal risks that the law could cause for attendees, aggressive rhetoric emerged during the bill's campaign that created a potentially hostile environment for gatherings of diverse individuals. Recognizing the threat to inclusivity and safety, conference organizers swiftly responded to the election results by surveying AAPS membership to evaluate potential responses to the exclusionary law (Newman 2016). Members' feedback indicated a strong desire to move the conference. A substantial number of respondents who had planned to attend the conference noted that they would not be comfortable doing so if the conference remained in North Carolina.

Overwhelmingly, AAPS committed to the organization's fundamental values and made the decision to move the conference from Raleigh, North Carolina, to St. Paul, Minnesota. The AAPS board even generated an open letter to the North Carolina governor that explained the decision to move the conference (BoD 2016). The letter's content was limited by the tax laws that prevent nonprofits from lobbying; the letter does not mention HB2 by name or call for its repeal (Pandya 2016). These conversations inspired AAPS to formalize the fundamental values for the organization after the board realized:

> we were acting on behalf of core values that we hold, but have yet to articulate as an organization—values of inclusivity, respect, and openness to all people regardless of race, ethnicity, class, age, religion, sexual orientation, gender identity, accessibility needs, or country of origin. This also made me realize the importance of more clearly defining our core values, even though we all know and embrace aspects of these values at some level—aspects that are at the very heart of [c*science] such as "anyone, anywhere can participate in science," "diversity and inclusivity," and being "a community of practice built on collaboration" that has as one of its goals "fostering diversity and inclusivity within the field."
>
> (Newman 2016)

In addition to formalizing the fundamental values of AAPS, membership discussions about appropriate responses provided the initial kindling for (1) the generation of a code of conduct for the 2019 conference that would eventually apply to all AAPS; (2) visioning actions for equity at conferences that yielded a strong environmental justice theme for 2019; (3) creative thinking to embed equity and justice into proposals and reviews for most symposia, workshops, and talks; and (4) opening virtual options for people that cannot physically or fiscally attend an in-person conference, including the employees of states boycotting travel to jurisdictions that criminalized bathroom choices. AAPS board members, general members, working groups, conference organizers, and local hosts accepted the compromises that were required to ensure all attendees were welcome at the 2017 conference venue.

The 2017 conference theme recognized the confluence of many tributaries into the mighty Mississippi River to represent the inclusion of diverse perspectives into AAPS. To overcome the mid-planning challenge of changing the venue, AAPS reevaluated membership and conference fees to ensure affordability and accessibility. This sets a pattern for AAPS to frequently update fee structures to reflect changes in context and membership from year to year. Conference organizers also partnered with the Science Museum of Minnesota, which reduced the planning time and effort required to coordinate the logistics for a public event held as part of the annual conference. Local partners around St. Paul, Minnesota, rallied in order to pull off the relocation without substantial compromises to the quality or availability of locally rooted content, partnerships, and field trips.

A striking aspect of the compromise to move the conference was the financial risk of relocation. AAPS has a financial responsibility to dedicate funds and assets toward the purpose of advancing the organizational mission, vision, goals, and objectives. The fiscal impacts of relocating presented a substantial risk to the fledgling organization. Specifically, AAPS contracts for the original 2017 conference totaled around $400,000, and cancellation fees would have totaled $143,162 if the conference was completely removed from North Carolina. For reference, the total operating budget for AAPS in the 2018 fiscal year was $83,468. The ability to move the conference was bolstered by hotels' willingness to endure the financial consequences of this boycott through their choice to forgive cancellation fees and allow rescheduling instead. The move of the AAPS conference, enabled by the flexibility of the venues and City of Raleigh, contributed to a $3.7 billion hit to North Carolina's economy due to

HB2 boycotts (Dalesio and Drew 2017). AAPS was able to make a moral and financial statement about HB2 by participating in this boycott without violating the lobbying regulations for nonprofits.

AAPS returned to North Carolina for the next conference after renegotiating Raleigh contracts in hopes that the state would have time to respond to the boycott with more just laws. Regrettably, HB2 was not adjusted during the two years between AAPS conferences beyond "in-name-only policy improvements [that] still leave many people marginalized and unsafe" (Shirk et al. 2018). Conference organizers and AAPS board members considered this potential outcome when the decision to relocate was originally made, as communicated in a 2016 call-to-action to members (Pandya 2016). The "existential risk" of continuing to boycott North Carolina that would lead to a severe financial hit to AAPS was explained transparently and publicly in a 2018 call-to-action to AAPS members (Shirk et al. 2018). For example, the conference rescheduling negotiations yielded an increase in conference hotel cancellation fees of up to $260,000 if AAPS fell short of the room quota. In addition to the financial realities, AAPS committed to return in 2019 to honor the powerful local partners that were committed to this conference and patiently waited for its return to their community (see the "Justice" section). Members, leadership, and conference organizers committed to implementing the return in a way that upheld the fundamental values of the organization.

With the decision to move back to North Carolina, conference organizers worked closely with LGBTQ+ organizations. By 2019, the City of Raleigh embraced a motto of "All Are Welcome Here" and built capacity to provide inclusive gathering venues that resisted or side-stepped the HB2 requirements (Visit Raleigh 2024). AAPS commitments from 2016 were fulfilled by switching tactics from the boycott to "partnering with local activists, highlighting the contributions LBGTQ[+] communities to [c*science], and promoting [c*science] that advances LBGTQ[+] priorities" (Pandya 2016). Further, AAPS faced both the moral and fiscal complexities of the venue relocation compromise by donating to the LGBT Center of Raleigh once conference attendees met room thresholds. Organizers proactively broadcasted this goal to attendees to help avoid hotel fees altogether—especially after AAPS encumbered $20,000 in fees for unfilled rooms in St. Paul due to 2015-era hotel contracts that were similar to those already locked in for 2019. Local LGBTQ+ activists were prominently featured in the conference lobby booths, which also provided attendees with personal opportunities to get help addressing their own navigation of Raleigh and North Carolina gender-related laws.

In 2018, AAPS understood that travel bans were still in place for potential attendees and unfortunately the organization could not fully honor those remaining boycott efforts. At the time of the 2019 return to Raleigh, AAPS considered investing in virtual participation to accommodate those that needed to boycott North Carolina but transparently chose to focus on facilitating in-person attendance to dig out of the financial reality of the rescheduling contracts. AAPS leadership specifically shared with membership that "we do not want to undermine the power and utility of such remaining boycotts, as we do not want to encourage (implicitly or otherwise) policies that we find abhorrent" (Shirk et al. 2018). Overall, the venue decisions summarized here underscore a commitment to inclusion and responsiveness that set a precedent for future conference planning and advocacy efforts. If the host organizations value the ability of attendees to participate comfortably as their authentic selves, conferences must be inclusive to people of all identities, but this pursuit is rarely straightforward.

Discussion and Reflection

The findings from over a decade of attempts to embed justice, equity, diversity, and inclusion principles into AAPS conferences reveal significant benefits for the c*science community that justified the investments required to make progress. The prioritization of these fundamental values enhanced the inclusivity of conferences and fostered broader engagement with diverse perspectives. Voices from historically excluded groups became more integral to the c*science dialogue as individual attendees could comfortably participate as their authentic selves and share their individual expertise. Historically excluded communities, environmental justice practitioners, and grassroots organizations built a crucial pillar that reinforced the AAPS mission. Consequently, the primary beneficiaries of these efforts were AAPS members, conference attendees, and the larger community of c*science practitioners. These beneficiaries gained enriched experiences, broader access to resources, platforms to share new developments, forums to work through challenges, and enhanced networking opportunities.

The AAPS was initially organized by predominantly white and women volunteers from academic, nonprofit, and government institutions. We acknowledge that many of these founders, including two of this paper's authors, collaborated with diverse communities and served as connectors to more inclusive networks. AAPS founders were often able to fulfill their institutions' requirements for academic service activities as a part of their employment, which provides space and time to volunteer for scientific associations. For example, the authors of this paper are white women who earn professional benefits in academia or nonprofits through publications and conferences. The efforts described in this chapter were often limited by a reliance on volunteer capacity, which is a luxury for most people beyond the immediate needs of their loved ones and neighbors. Scientific organizations must carefully consider and mitigate the exclusion of those without professional rewards for service from their participation in conference planning and attendance.

The adaptive strategies used by AAPS to plan conferences highlight the potential for other scientific communities to rethink their conference models, such as relocating events to ensure safety and inclusiveness or restructuring formats to enhance accessibility. This study emphasized the importance of frequent adaptation and commitment to fundamental values as a challenge for scientific societies to reflect on their inclusivity and equity practices. New questions that emerge from this study include how other scientific associations can effectively integrate community perspectives and maintain long-term commitment to justice, equity, diversity, and inclusion values. AAPS plans to continually build on these findings by expanding outreach efforts, increasing diversity in leadership, and exploring new ways to support community-driven research and action.

The study's potential significance extends beyond the AAPS and offers a model for other societies and organizations that aim to embed justice, equity, diversity, and inclusion principles into their operations. By prioritizing justice and equity, these organizations can improve their relevance to current events, the representativeness of their knowledge base, and the impact of their research beyond the bounds of their own discipline. By prioritizing inclusivity and diversity, these organizations can create forums that benefit a wider range of participants and ultimately enrich the collective knowledge base. New questions that emerge from this study include how to sustainably maintain and further these gains, what additional strategies can be employed to overcome remaining barriers to full inclusivity, and how to measure the long-term impacts of these efforts especially when metrics of success vary substantially across community contexts. To continue this inquiry, we will conduct longitudinal studies to track the progress of justice, equity, diversity, and inclusion initiatives within AAPS and similar organizations. Additionally, we will explore qualitative research methods to capture the nuanced experiences of diverse participants and identify areas for further improvement. Continuous dialogue among attendees, members, hosts, conference organizers, and sponsors is foundational to AAPS conference planning and implementation. We aim to refine our approaches and ensure that our practices remain responsive to evolving needs and challenges in the field of c*sciences.

Discussion Questions

These questions are designed to prompt reflection and action among readers, encouraging them to consider how the experiences and strategies of AAPS might inform their own practices and the broader scientific community's approach to justice, equity, diversity, and inclusion.

1. *Applicability to other contexts*—How can scientific communities ensure that their efforts toward justice, equity, diversity, and inclusion are sustainable and not just temporary initiatives?
2. *Sustaining commitments*—What mechanisms can be put in place to measure the long-term impact of justice, equity, diversity, and inclusion initiatives in scientific organizations?
3. *Balancing tradition and innovation*—How can organizations balance the inclusion of diverse voices with the potential for creating echo chambers within their networks?
4. *Long-term impacts*—What additional strategies can be employed to overcome barriers to increase participation for historically excluded groups in scientific conferences?
5. *Building inclusive networks*—How can lessons learned from the AAPS experience be adapted and implemented in other professional and academic settings to enhance inclusivity and equity?

Conclusions

AAPS consistently and sustainably invested in justice, equity, diversity, and inclusion over a decade since the foundational 2012 conference. The conferences were forums to attempt new directions, test new initiatives, and broadcast innovations within the c*science field related to these fundamental values. AAPS navigated substantial changes in terrain over the past decade: conditions changed, membership evolved, and perspectives expanded. Very few elements remained static from conference to conference. AAPS members and conference attendees were open to these new experiences. While not every effort was successful, the nimbleness of conference organizers and constructive feedback from attendees were highly valued when pivoting to new directions and initiatives. Additionally, the attendees and members that were frustrated with the lack of progress or the abruptness of a change were provided with opportunities to provide genuine criticism and calls-to-action for change. AAPS leaders and conference organizers found that transparency, welcoming discussion forums, and creative solutions were critical to rebound from mistakes. For example, compromises like alternating between virtual and in-person events helped to avoid unilateral decisions and to equitably distribute conference benefits. The efforts described in this chapter did not create a perfect recipe for justice, equity, diversity, and inclusion that could be replicated with ease. Instead, AAPS's collective efforts to improve the conference helped us develop a playbook; some plays are proactive, like pre-conference membership surveys, and others are reactive to respond to new challenges, like a global pandemic. This playbook was optimized and is still used by skilled and trained facilitators within the AAPS leadership and working groups. Please be reminded that practice is required before the big day of the conference arrives and that multiple attempts will be required to accomplish complicated goals like justice, equity, diversity, and inclusion.

Achieving authentic progress in justice, diversity, equity, and inclusion requires continuous commitment where each member and leader is held accountable to these fundamental values. This can be accomplished through an integrative approach that includes implementing a robust code of conduct; requiring presenters to disclose justice, diversity, equity, and inclusion information in their abstract submissions and presentations alongside their conflicts of interest; and celebrating both innovative approaches and traditional success stories that advance these fundamental values. Scientific associations must prioritize the fundamental values of justice, equity, diversity, and inclusion when planning conferences. By embedding these principles into every aspect of conference planning, associations can ensure that they are not only meeting, but exceeding, their obligations to foster a just and inclusive environment.

Note

1 The authors thank the AAPS board; AAPS director; Justice, Equity, Diversity, and Inclusion Working Group; and Environmental Justice Working Group for their contributions to the efforts described here. We also thank Darya Chernikhova for their editorial support.

We respectfully acknowledge the Indigenous and First Nations peoples as the original custodians of the lands on which we live and work. We honor their enduring connection to the land, waters, and skies and recognize their deep knowledge and wisdom that has been passed down through generations. In the context of participatory sciences, we pay tribute to the invaluable contributions of Indigenous communities who have practiced forms of community science for millennia, grounded in a profound understanding of the natural world.

This chapter on justice, diversity, equity, and inclusion in participatory sciences is inspired by and indebted to the rich traditions and practices of Indigenous peoples. We recognize that true inclusion in participatory sciences requires not only the inclusion of diverse voices but also the integration of Indigenous knowledge and wisdom, which challenge and enrich our scientific endeavors.

We express our gratitude to the Indigenous elders, knowledge holders, and community members who continue to guide us in our efforts to create spaces that are truly inclusive, equitable, and just. May this work honor your legacy and contribute to a future where participatory sciences reflect the values, wisdom, and practices of all communities.

References

AAPS (Association for Advancing Participatory Sciences). 2015a. "Citizen Science 2015: Inaugural Conference of the Citizen Science Association Conference Program." San Jose, CA. https://participatorysciences.org/wp-content/uploads/2016/07/cs_confprogrm.pdf.

AAPS (Association for Advancing Participatory Sciences). 2015b. "Citizen Science 2015 Conference Home Page." https://participatorysciences.org/conferences/citsci2015/.

AAPS (Association for Advancing Participatory Sciences). 2015c. "Presenters List for Citizen Science 2015." https://participatorysciences.org/conferences/citsci2015/presenters-list-for-citizen-science-2015/.

AAPS (Association for Advancing Participatory Sciences). 2019. "CitSci2019 Homepage." https://participatorysciences.org/conferences/citsci2019/.

AAPS (Association for Advancing Participatory Sciences). 2021a. "Code of Conduct for AAPS Events." Policy version 2. Last modified February 18, 2021. https://participatorysciences.org/code-of-conduct/.

AAPS (Association for Advancing Participatory Sciences). 2021b. "CitSciVirtual 2021 Home Page." https://participatorysciences.org/conferences/citscivirtual/.

AAPS (Association for Advancing Participatory Sciences). 2022. "C*Sci 2022 Homepage." https://participatorysciences.org/conferences/c-sci-2022/.

AAPS (Association for Advancing Participatory Sciences). 2024. "About the Association for Advancing Participatory Sciences." https://participatorysciences.org/about/.

Benz, S., A. Miller-Rushing, M. Domeroese, H. Ballard, R. Bonney, T. DeFalco, S. Newman, J. Shirk, and A. Young. 2013. "Workshop 1: Conference on Public Participation in Scientific Research 2012: An International, Interdisciplinary Conference." *Bulletin of the Ecological Society of America* 94 (1): 112–17. https://doi.org/10.1890/0012-9623-94.1.112.

BoD (Board of Directors). 2016. "Open Letter to the Honorable Patrick McCrory." Citizen Science Association, October 20. https://participatorysciences.org/wp-content/uploads/2016/10/Citizen-Science-Association-Board-HB2-Letter-20-OCT-2016.pdf.

Chew Bigby, B., E. Hatley, and R. Jim. 2022. "The Potential of Toxic Tours: Indigenous Perspectives on Crises, Relationships, Justice and Resurgence in Oklahoma Indian Country." *Journal of Sustainable Tourism* 31 (12): 2645–66. https://doi.org/10.1080/09669582.2022.2116642.

Cooper, C. B., C. L. Hawn, L. R. Larson, J. K. Parrish, G. Bowser, D. Cavalier, R. R. Dunn, M. (Muki) Haklay, K. Kar Gupta, N. O. Jelks, V. A. Johnson, M. Katti, Z. Leggett, O. R. Wilson, and S. Wilson. 2021. "Inclusion in Citizen Science: The Conundrum of Rebranding." *Science* 372 (6549): 1386–8. https://doi.org/10.1126/science.abi6487.

Dalesio, E.P., and J. Drew. 2017. "'Bathroom Bill' to Cost North Carolina $3.76B." *AP® News*, March 30. https://apnews.com/article/e6c7a15d2e16452c8dcbc2756fd67b44.

Duncan, S., A. Cawood, J. Tuttle, and R. Putnam. 2024. "Organizing Facilitated Poster Discussion Sessions: Their Inspiration, Planning, and Implementation at the C*Sci2023 Conference." *Citizen Science: Theory and Practice* 9 (1): 16. https://doi.org/10.5334/cstp.716.

Edelheim, J. R., K. Thomas, K. G. Åberg, and G. Phi. 2017. "What Do Conferences Do? What Is Academics' Intangible Return on Investment (ROI) from Attending an Academic Tourism Conference?" *Journal of Teaching in Travel and Tourism* 18 (1): 94–107. https://doi.org/10.1080/15313220.2017.1407517.

Eitzel, M. V., J. L. Cappadonna, C. Santos-Lang, R. E. Duerr, A. Virapongse, S. E. West, C. C. M. Kyba, A. Bowser, C. B. Cooper, A. Sforzi, A. N. Metcalfe, E. S. Harris, M. Thiel, M. Haklay, L. Ponciano, J. Roche, L. Ceccaroni, F. M. Shilling, D. Dörler, F. Heigl, T. Kiessling, B. Y. Davis, and Q. Jiang. 2017. "Citizen Science Terminology Matters: Exploring Key Terms." *Citizen Science: Theory and Practice* 2 (1): 1. https://doi.org/10.5334/cstp.96.

Ellwood, E. R., G. B. Pauly, J. Ahn, K. Golembiewski, L. M. Higgins, M. A. Ordeñana, and M. von Konrat. 2023. "Citizen Science Needs a Name Change." *Trends in Ecology and Evolution* 38 (6): 485–9. https://doi.org/10.1016/j.tree.2023.03.003.

Etzion, D., J. Gehman, and G. F. Davis. 2022. "Reimagining Academic Conferences: Toward a Federated Model of Conferencing." *Management Learning* 53 (2): 350–62. https://doi.org/10.1177/13505076211019529.

Geiser, K., and G. Waneck. 1983. "PCBs and Warren County." *Science for the People* 15 (4): 13–17.

Heaney, C. D., S. M. Wilson, and O. R. Wilson. 2007. "The West End Revitalization Association's Community-Owned and -Managed Research Model: Development, Implementation, and Action." *Progress in Community Health Partnerships: Research, Education, and Action* 1 (4): 339–49. https://doi.org/10.1353/cpr.2007.0037.

Heimlich, J. E. 2012. "Public Participation in Scientific Research: 2012 Conference Evaluation." *Reimagining Equity and Values in Informal STEM Education*, September 10. https://informalscience.org/wp-content/uploads/2019/02/Final_PPSR_Conference_2012_Evaluation_Report.pdf.

Heimlich, J. E., and G. Timko. 2015. "Citizen Science Conference Full Evaluation Report." *Reimagining Equity and Values in Informal STEM Education*, August 1. https://informalscience.org/evaluation/citizen-science-conference-full-evaluation-report/.

Hils, A., S. Kelty, A. Hjarding, L. Silka, and A. Cawood. 2023. "Learning from the First Citizen Science Association Virtual Conference." *Citizen Science: Theory and Practice* 8 (1): 17. https://doi.org/10.5334/cstp.559.

Kelty, S., S. Kafka, G. L. Jenni, C. Cooper, L. Rasmussen, O. Wilson, V. Bitting, V. Martin, S. Wilson, A. Wilson, A. Hjarding, J. Oliver, M. Bear, M. V. Eitzel, L. Higgins, S. Tayag, and other C*Science Practitioners. 2022. "Harmonized JEDI Definitions for C*Science." *Association for Advancing Participatory Sciences*. https://participatorysciences.org/harmonized-jedi-definitions/.

Newman, G. 2016. "In Support of Our Core Values: A Conference Decision." *Association for Advancing Participatory Sciences*, May 26. https://participatorysciences.org/2016/05/26/in-support-of-our-core-values-a-conference-decision/.

Pandya, R. 2016. "Setting the Table: An Open Letter to North Carolina." *Association for Advancing Participatory Sciences*, October 20. https://participatorysciences.org/2016/10/20/setting-the-table-an-open-letter-to-north-carolina/.

Putnam, R. 2023. "Announcing a New Name for This Association." *Association for Advancing Participatory Sciences*, July 14. https://participatorysciences.org/2023/07/14/announcing-a-new-name-for-this-association/.

Shirk, J. L. 2022a. "Coming to Terms." *Association for Advancing Participatory Sciences*, May 5. https://participatorysciences.org/2022/05/05/coming-to-terms/.

Shirk, J. L. 2022b. "Coming to Terms: Exploring Language and Practices." *Association for Advancing Participatory Sciences*, December 30. https://participatorysciences.org/2022/12/30/coming-to-terms-language-and-practices/.

Shirk, J. L. 2024. "Reintroducing This Association: Advancing Participatory Sciences." *Association for Advancing Participatory Sciences*, April 2. https://participatorysciences.org/2024/04/02/reintroducing-this-association-advancing-participatory-sciences/.

Shirk, J., C. Cooper, and A. Bowser. 2018. "On Showing Up and Standing Up: Holding the CitSci2019 Conference in Raleigh, North Carolina." *Association for Advancing Participatory Sciences*, August 21. https://participatorysciences.org/2018/08/21/on-showing-up-and-standing-up-holding-the-citsci2019-conference-in-raleigh-north-carolina/.

Shirk, J. L., H. L. Ballard, C. C. Wilderman, T. Phillips, A. Wiggins, R. Jordan, E. McCallie, M. Minarchek, B. V. Lewenstein, M. E. Krasny, and R. Bonney. 2012. "Public Participation in Scientific Research: A Framework for Deliberate Design." *Ecology and Society* 17 (2): 29. http://www.jstor.org/stable/26269051.

Visit Raleigh. 2024. "All Are Welcome Here." https://www.visitraleigh.com/plan-a-trip/raleigh-pride/all-are-welcome-here/.

Wilson, S., moderator. 2019. "CitSci2019 Environmental Justice Panel." Association for Advancing Participatory Sciences, March 22. YouTube. Video, 1 hr., 48 min., 48 sec. https://www.youtube.com/watch?v=5FL2bCifB_A.

10

Lessons in Undergraduate Biology Education: Navigating Adaptive Challenges Using an Evolving Community of Practice

Jaclyn Reeves-Pepin, Vedham Karpakakunjaram, Kirstin J. Milks, Anna Hiatt, and Stacey Kiser

Professionals in academia, especially in higher education, expect their discipline-based professional societies to serve multiple critical roles, including advisory, professional validation, mentorship, and leadership.[1] Since its inception in 1938 (Figure 10.1, marker 1), the National Association of Biology Teachers (NABT) has been instrumental in filling these roles for biology educators across North America, modifying its organizational framework and efforts to fit the changing landscape of biology education and to meet the changing requirements of instructors (Holmberg et al. 2021). In turn, NABT has played a critical role in shaping the quality and standards of biology curriculum and education along the kindergarten to postsecondary education (K–16) continuum.

Better engagement between professional societies and faculty at community colleges and two-year institutions, in particular, has long been recognized as a key aspect of improving diversity, equity, and inclusion in STEM disciplines and organizations. Professional societies "benefit from intentional and systematic actions to include community college faculty" (Maier et al. 2021), as these educators bring boots-on-the-ground knowledge of curriculum, instruction, and the rapidly changing needs of their communities and students. The benefit derived from community college instructors engaging with disciplinary societies is mutual and counters the isolation experienced by many community college faculty (Bickerstaff 2024).

Considering that 53 percent of Indigenous American, 48 percent of Hispanic, 39 percent of Black, and 36 percent of first-generation students in the United States who have earned an undergraduate degree have taken courses at a community college, the two-year setting plays a critical role in diversifying higher education (AACC 2024, 2). Engaging and leveraging the knowledge of two-year faculty to build and deliver inclusive, high-quality science education on the national scale is an essential pathway to increasing diversity in the nation's STEM communities (Labov 2017). Yet barriers persist for two-year instructors to contribute to change-making initiatives in STEM education. This includes a persistent emphasis on funding quantitative and well-controlled research in undergraduate curriculum and instruction that can be challenging to implement and support at many two-year institutions.

A Community of Practice Emerges, Spanning Professional Learning Communities

As of 2024, 14 percent of NABT members report teaching at two-year institutions, while 22 percent teach at four-year colleges and universities and 50 percent teach at the high school level (grades 9–12). This case study describes an emergent shift within our organization, as established professional learning communities (PLCs) centered on member context (high school, two-year, or four-year undergraduate-serving institutions) began to cross-pollinate, creating a new community of practice focused on recentering introductory, college-level, biology instruction on authentic scientific practice.

We define PLCs as groups engaged actively and collaboratively in professional learning with the objective of growing professional identities, expanding knowledge, and improving pedagogical strategies (IES 2019; Lee and Shari 2012). In contrast, a community of practice is a team of people working together with a common goal of addressing a concern or passion around a theme (Wenger et al. 2002), and it has been applied as a critical part of the change theory that can bring about system changes to STEM education (Reinholz et al. 2021; Reinholz and Andrews 2024). Communities of practice have three distinct characteristics: (1) domain, meaning a common purpose and value; (2) a community; and (3) a practice that comprises the knowledge, frameworks, and resources shared by that community (Kezar and Gehrke 2015). Wenger et al. (2002) further describe that communities of practice actively engage in the learning process through social participation in shared practices, through which learning occurs inherently.

While communities of practice share central tenets of PLCs, they are also a distinct class of system. In practice, PLCs tend to be intentionally created with a defined hierarchy (Kezar and Gehrke 2015) and focus on specific topics and predetermined goals, while communities of practice are more organic, less stratified, and tend to emerge and evolve in response to changing goals. This does not make communities of practice a less rigorous framework for generating knowledge; rather, as this case study illustrates, they can connect holders of knowledge across contexts and expertise, allowing nimble, coordinated, and just-in-time creative responses to multifaceted challenges and even paving the way for future PLC-based efforts.

NABT was first founded in 1938 to support secondary science teaching. As it grew to include faculty in higher education, the society was divided by geographical region in the early 1960s to facilitate more effective communication and community building (Jeffers and Lightner 2001). This organizational strategy then shifted again as an increasing number of community college educators began joining NABT during the baby boomer-fueled educational expansion of the late 1960s and 1970s. Community college members successfully petitioned for a change to the association's bylaws that formed a specific PLC; in 1983, the NABT Two-year College Section was established with the express purpose of creating programs to specifically support community college educators. NABT eventually formalized additional sections for educators, and in addition to the two-year college level, NABT currently supports sections for teachers at the high school level, Advanced Placement teachers, four-year colleges and universities, and informal educators (Figure 10.1, marker 2). These sections function as PLCs within the larger organization to help design and support programs, resources, awards, and policies for their teaching level-specific constituencies.

The sections also provide myriad volunteer opportunities that help members grow and expand their professional network as well as raise their profile at a national level. These pathways for career development are especially important for high school and two-year college instructors (both full-time and adjunct) who may otherwise have limited opportunities within their school districts and college systems.

While the sections continue to connect around problems of practice specific to the learners they serve, they also intersect in ways that encourage community members to interact dynamically across section PLCs and within the larger NABT network through programs like the NABT national conference; the association's peer-reviewed journal, *The American Biology Teacher;* and additional programmatic offerings throughout the year.

Introductory Biology Emerges as a National Priority in Science Education

Between 2009 and 2013, multiple national efforts began to develop new frameworks for biology education across K–16 with a focus on competencies and the integration of concepts and skills (Figure 10.1, markers 4 and 5). The National Research Council publicly released *A Framework for K–12 Science Education* in 2011 (NRC 2012), the basis for the *Next Generation Science Standards* (NGSS) for K–12 that were released in 2013. Also in 2011, the College Board updated its *Advanced Placement Biology Course and Exam Description* for college-level introductory biology in the high school setting (College Board 2011). At the collegiate level, the American Association for the Advancement of Science released the *Vision and Change in Undergraduate Biology: A Call to Action* report (hereafter referred to as *Vision and Change*) in 2011 (AAAS 2011).

NABT members were integral in the development of these national initiatives, and the organization helped identify and convene subject matter experts for all three reports during the draft phases. For undergraduate biology specifically, NABT contributions included, but were not limited to: the full NABT executive committee attended the national conference marking the release of *Vision and Change* in 2009; in 2012 a full-day Faculty Development Summit focused on implementing *Vision and Change* was hosted at the NABT national conference; NABT served as a host for the Biology Directors' Consortium, providing space for discussing the impact of *Vision and Change* on large-enrollment biology course; and educators and researchers from NABT were both leaders and participants in the Partnership for Undergraduate Life Sciences Education (PULSE) network that focused on institutional mechanisms of change.[2]

At the same time that undergraduate biology was being influenced by *Vision and Change,* several NABT members were involved in the College Board's revision of the AP® Biology course and exam. This updated standards framework, like the NGSS and *Vision and Change,* required instructors to simultaneously teach science content and science practices. NABT partnered with the Biological Sciences Curriculum Study (now, BSCS Science Learning) and the Howard Hughes Medical Institute (HHMI) to create a multi-year leadership academy to help AP® Biology teachers update and share curriculum and instructional strategies aligned with the new AP® framework. NABT, therefore, was equally prepared to support AP® Biology teachers prior to the release of their new framework.

This programming that arose from the NABT members of AP®, Two-year and Four-year Sections had a common thread: undergraduate-level introductory biology was being affected by multiple national initiatives that were going to impact AP®, two-year, and four-year biology teachers across the country. Recognizing the synergy occurring across the different initiatives, the NABT Board of Directors—with the support from the Introductory Biology Project, a National Science Foundation-funded Research Coordination Network led by NABT past president Dr. Gordon Uno—established the Introductory Biology Task Force as an ad hoc committee in 2014 to bring together leaders from all three sections to address the unique challenges of teaching and learning undergraduate introductory biology (Figure 10.1, marker 6).

Collaboration between Multi-level Educators in Transformational Efforts

The initial Introductory Biology Task Force was a community of practice that assembled several NABT leaders: past presidents; sitting members of the board of directors; section leadership from the AP® Biology, two-year, and four-year levels; and stakeholders from the College Board and Educational Testing Services. Early brainstorming sessions with the task force focused on the alignment between *Vision and Change*, the NGSS, and the AP® framework, and resulted in a well-attended conference session at the NABT national conference in which participants collaborated to inform an updated toolkit for teachers of introductory biology across contexts.[3]

Subsequent NABT conference sessions in 2015, 2016, and 2017 were designed to help introductory biology teachers across contexts unpack different content areas and skills to integrate and map into their curricula. Learning from these sessions continues to inform the community of practice within NABT, with at least three teams at the 2024 NABT annual conference incorporating alignment between the frameworks as part of their presentations on helping students at multiple levels craft explanations and arguments, building authentic assessment practices for life science courses, and centering biology instruction around the United Nations Sustainable Design Goals.

The learning generated by the NABT community of practice extended beyond the organization as well. Multiple members of the Introductory Biology Task Force attended the inaugural Gordon Research Conference on Undergraduate

Biology Education in 2015 and again in 2017; at the latter, task force members presented a poster on the alignment of the frameworks and how the alignment can be used to inform introductory biology curricula. That same year, the NABT executive director presented an introduction to the task force and an update on ongoing work to the National Academies of Sciences, Engineering, and Medicine Board on Life Science. The task force also submitted several grant proposals to the NSF and in 2018 was awarded an Improving Undergraduate STEM Education (IUSE) conference grant. The Advancing Introductory Biology conference was held in August 2019 at the HHMI conference center and included a total of sixty-seven participants (Figure 10.1, marker 7). The instructor attendees comprised 60 percent four-year college or university faculty, 23 percent two-year college faculty, and 17 percent high school and AP® Biology faculty. A small portion of attendees taught at multiple institution types (e.g., taught as an adjunct at both a four-year and two-year school or taught at both a high school and a two-year school). Organizers noted in their final report that several attendees highlighted the positive impact that cross-institutional collaborations between two-year, four-year, and high school instructors had on their experience. Five years later, this observation continues to be made in post-event surveys conducted by NABT.

Collaborations and network building among this diverse group of classroom practitioners helped increase interest in both national and regional communities of practice around introductory biology. One outcome of the Advancing Introductory Biology conference was the formal establishment of the Introductory Biology Experience (IBEx), a new PLC composed of task force leadership, regional leaders of the NABT, and partners from other projects (Figure 10.1, marker 7). Supported by the NSF-supported Faculty Developers Network, IBEx members convened in December 2019 to develop regional networks and cross-level professional development based on the previous work of the Introductory Biology Task Force. The work was interrupted by the Covid-19 pandemic, forcing the PLC to pivot to an online experience. In August 2020, NABT hosted a three-day event entitled Community Conversations: The Introductory Biology Experience (Figure 10.1, marker 8). Three guest speakers were invited to lead a two-hour interactive discussion with seventy-six participants about effective online teaching tools, inclusive teaching practices, and how to teach science practice skills in an online environment. The Community Conversations series has continued to thrive, and webinars designed to attract a mixed audience of instructors from high schools, two-year colleges, and four-year colleges and universities are now held multiple times a year.

The Introductory Biology Task Force as a Force Multiplier for Other Projects

For the last few years, the Introductory Biology Task Force has continued to seek partnerships and provide a supportive platform to bolster PLCs around the introductory biology course. One specific example from the two-year setting is the NSF-supported Quantitative Biology@Community Colleges (QB@CC) project launched in 2019 (Figure 10.1, marker 7). Building from a workgroup supported by the National Institute for Modeling Biological Systems (NIMBioS; formerly the National Institute for Mathematical and Biological Synthesis), the project surveyed two-year faculty to identify barriers to teaching quantitative skills in the biology curriculum. Responses indicated predictable impediments such as lack of institutional support, finding substitutes to cover classes, and appropriate reward systems for engagement with professional development opportunities (Brownell and Tanner 2012; Corwin et al. 2019). In addition, the lack of pedagogical content knowledge emerged as a critical challenge to integrating quantitative skills in the biology curriculum (Corwin et al. 2019).

Creating a PLC to address this pedagogical content knowledge barrier was one of the main goals of the QB@CC project. Both the principal investigator (PI) and co-PI for QB@CC were participants in the Advancing Introductory Biology conference held in 2019, and they were able to leverage the existing network to make early connections with other two-year faculty and partnering organizations to support this new initiative. NABT, and specifically its Two-year College Section, also provided critical support for the QB@CC project. The QB@CC leadership team shared project opportunities at the 2019 and 2020 NABT annual conferences, recruiting participants to their network. Starting in 2021, participants were able to disseminate curricular modules and other products from the QB@CC project during sessions at the NABT conference (Esquibel et al. 2023). In September 2023, QB@CC and NABT co-hosted a two-hour webinar highlighting a popular learning module; over 130 people registered for the webinar with 36 percent of the final attendees being two-year college faculty, 27 percent being faculty from four-year colleges and universities, and 33 percent being from the high school level (Figure 10.1, marker 9).

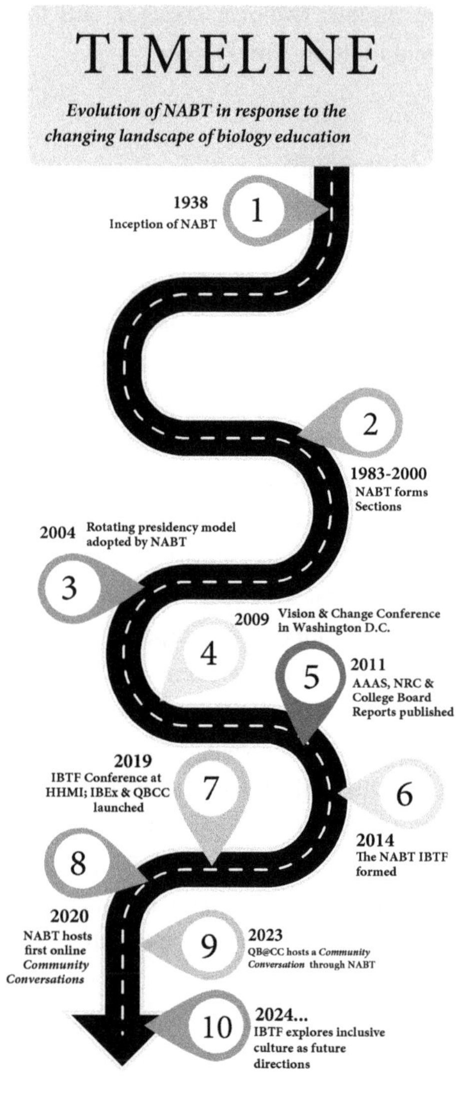

Figure 10.1 A roadmap that indicates landmark transformations in the National Association of Biology Teachers (NABT) membership and leadership structures since its inception over 80 years ago. Since 2004, when the rotating presidency model was adopted, NABT has been a hub for creating transformational experiences for biology educators. The figure highlights two major landmarks—launch and sustained impact of Introductory Biology Task Force (IBTF) on K–16 biology education and Quantitative Biology@Community Colleges (QB@CC) as a professional space for community college faculty. (AAAS = American Association for the Advancement of Science; HHMI = Howard Hughes Medical Institute; IBEx = Introductory Biology Experience; NRC = National Research Council) Created in Canva by Vedham Karpakakunjaram and Jaclyn Reeves-Pepin.

By engaging with the established network initially developed by the Introductory Biology Task Force, the two-year faculty involved in the QB@CC project were able to showcase their curricular materials and classroom scholarship to a broad audience of fellow educators.

Lessons in Diverse Leadership to Face Adaptive Challenges

The Introductory Biology Task Force is just one example of how NABT's integrated PLCs have served as an incubator for a community of practice that encourages inclusive leadership and engages in national efforts that benefit diverse educators from multiple institutional settings. It also provides an instructive context for professional organizations interested in building capacity in science education to address adaptive challenges.

First defined by Ronald A. Heifetz (2009) at Harvard University's Kennedy School of Government, "adaptive challenges" are complex, unknown to the group, and require new learning to be addressed. While technical challenges are often complicated and solving them is critically important, the necessary knowledge exists to enact a solution. Heifetz often uses the example of "replacing a faulty heart valve during cardiac surgery" as a technical challenge.

Adaptive challenges, on the other hand, "can only be addressed through changes in people's priorities, beliefs, habits, and loyalties" (Heifetz 2009, 19). These challenges transform groups by disrupting established behaviors and systems. It is important to note that adapting to new environments also generates loss. Some ingrained systems and cultural norms that made something (e.g., an undergraduate science course) successful in the past may need to be abandoned or rearranged so new systems can be put in place.

Educational settings are especially prone to challenges that have components that are both technical and adaptive. This is due to a number of complex and interconnected factors, including changes in the population (both demographic and population size), shifting political and professional priorities, and unpredictable funding.

Transforming the introductory biology course to improve outcomes for learners holding underrepresented identities or living in underserved communities is an adaptive challenge that required a new community of practice to emerge to help address it. Some organizational-level lessons from the experience with the Introductory Biology Task Force are presented below.

Encourage Bottom-up, Inclusive, Networked, Improvement Efforts alongside Top-down Initiatives

Teaching faculty in two-year settings have the same opportunities as those in four-year and high school settings to prototype and iterate novel instructional approaches in real time. In the case of the Introductory Biology Task Force, this access to opportunity allowed for deep and meaningful participation by educators serving a broad range of students and communities. This approach differs from one focused on more quantitative education research where barriers to inclusive participation may include low institutional support, limited access to funding and local research infrastructure, and lack of release time beyond the fundamental duties of an assigned teaching load.

Ensure Programs and Pathways Offer Professional Currency That Is Aligned with the Needs and Interests of the Target Audience(s) of the Work

For biology and education researchers, funded grants and accepted peer-reviewed publications serve as important professional currency: they bring legitimacy to research, pave the way for future endeavors, and are recognized and rewarded by home institutions. However, for professionals whose roles are more directly focused on classroom instruction, new educational resources and novel curriculum and instruction are a much more important professional currency, both to educator identity and in the eyes of their home institutions.

Additionally, many two-year and four-year members of NABT anecdotally report that it can be challenging to connect positively and productively with other educators at their home institutions around the work of improving science education, especially in the context of building diversity, equity, and inclusion. Experiences like the Introductory Biology Task Force can meet the important need for social connection and solidarity for educators who otherwise might not feel they have the personal bandwidth to sustainably engage in this essential work.

Build Authentic Leadership Pathways within the Organization for Professionals Engaged in Education and Outreach That Aims to Diversify the STEM Workforce

NABT hosts a portfolio of opportunities for classroom-based educators to professionally leverage their learning from qualitative and design-based inquiry into their practice. Currently, these include submitting a proposal for the annual conference, writing for the peer-reviewed journal, presenting at online events, becoming active in the organization's PLCs and emerging communities of practice, and serving as a volunteer on one of its many committees. By intentionally designing these opportunities, NABT has been able to build and maintain leadership capacity throughout all levels of the organization. This is significant because, beginning in 2004, NABT has supported a rotating presidency with individuals from the K–12 level being followed by a four-year representative, followed by a two-year representative (Figure 10.1, marker 3). This pattern ensures NABT has at least one executive officer serving, and advocating for, each educational level. Shared and equal leadership has allowed NABT to develop and maintain a collaborative culture that exerts a positive influence on biology education across the entire K–16 continuum.

Conclusion

For over eight decades, NABT has been able to mobilize its biology education community to grapple with complex challenges, leverage new opportunities when they arise, experiment, take risks, assist partners in mission-centered work, and thrive. The multi-year impact of the Introductory Biology Task Force is not only a case study on how a community of practice was formed to improve the undergraduate biology course; it is a testament to the organization's ongoing commitment to empower educators, encourage students, and help build a more diverse and inclusive STEM community.

Notes

1 The authors would like to thank Gordon Uno, whose initial support and continued involvement with improving "intro bio" has been instrumental in this project.

They would also like to express their gratitude to Dennis Gathmann, NABT's Two-Year Section historian, whose recordkeeping helped us to document organizational transformations with accuracy; Brad Williamson (2002 NABT president) and Bob Melton (2016 NABT president), who shared their accounts of leadership experiences and anecdotes of NABT's values that shaped their professional career; and our NABT membership, for their commitment to innovation that reimagines inclusive excellence in biology education.

2 More information on the PULSE Network can be found at https://pulse-community.org/.

3 The product of the work done by the NABT Introductory Task Force is available at https://nabt.org/Intro-Bio-Task-Force.

References

AAAS (American Association for Advancement of Science). 2011. *Vision and Change in Undergraduate Biology Education: A Call to Action*. AAAS. https://www.aaas.org/sites/default/files/content_files/VC_report.pdf.

AACC (American Association of Community Colleges). 2024. "Fast Facts 2024." https://www.aacc.nche.edu/research-trends/fast-facts/.

Bickerstaff, S. 2024. "Building Stronger Collaborations between Community Colleges and Disciplinary Societies." *The Mixed Methods Blog*, February 26. https://ccrc.tc.columbia.edu/easyblog/building-stronger-collaborations-between-community-colleges-and-disciplinary-societies.html.

Brownell, S. E., and K. D. Tanner. 2012. "Barriers to Faculty Pedagogical Change: Lack of Training, Time, Incentives, and … Tensions with Professional Identity?" *Cell Biology Education–Life Sciences Education* 11 (4): 339–46. https://doi.org/10.1187/cbe.12-09-0163.

College Board. 2011. *AP® Biology Curriculum Framework: 2012–2013*. College Board.

Corwin, L. A., S. Kiser, S. M. LoRe, J. M. Miller, and M. L. Aikens, M. L. 2019. "Community College Instructors' Perceptions of Constraints and Affordances Related to Teaching Quantitative Biology Skills and Concepts." *Cell Biology Education–Life Sciences Education* 18 (4): ar64. https://doi.org/10.1187/cbe.19-01-0003.

Esquibel, J., D. L. Rook, S. M. LoRe, J. H. Starnes, J. M. Miller, J. G. Buntz, A. Hugo, C. B. Nieuwsma, H. Seitz, A. Bissell, L. Gross, S. Kiser, S. Lenhart, M. A. Mills, C. Neuhauser, I. Corriette, S. Prescott, K. P. Jenkins, and V. Karpakakunjaram. 2023. "Quantitative Biology at Community Colleges, a Network of Biology and Mathematics Faculty Focused on Improving Numerical and Quantitative Skills of Students." *Cell Biology Education–Life Sciences Education* 22 (2): ar16. https://doi.org/10.1187/cbe.21-09-0244.

Heifetz, R. A., A. Grashow, and M. Linsky. 2009. *The Practice of Adaptive Leadership: Tools and Tactics for Changing Your Organization and the World*. Harvard Business Press.

Holmberg, T. J., S. Gusky, S. Kiser, V. Karpakakunjaram, H. Seitz, L. Fletcher, L. Fields, A. Nenortas, A. Corless, and K. Marcos. 2021. "Biology Educators, Professional Societies, and Practitioner Networks within Community Colleges." *New Directions for Community Colleges* 2021 (194): 15–28. https://doi.org/10.1002/cc.20449.

IES (Institute of Education Sciences). 2019. *Professional Learning Communities Facilitator's Guide for the What Works Clearinghouse Practice Guide*. https://ies.ed.gov/ncee/edlabs/regions/southeast/plc.asp.

Jeffers, G., and J. Lightner. 2001. *As We Remember It: A Personal Recollection of the First Four Decades of the National Association of Biology Teachers*. JLX Publishers.

Kezar, A., and S. Gehrke. 2015. "Communities of Transformation and Their Work Scaling STEM Reform." In *Pullias Center for Higher Education*. Rossier School of Education, University of South Carolina. Accessed March 30, 2021. https://pullias.usc.edu/download/communities-of-transformation-and-their-work-scaling-stem-reform/.

Labov, J. 2017. "Changing and Evolving Relationships between Two- and Four-Year Colleges and Universities: They're Not Your Parents' Community Colleges Anymore." *Cell Biology Education–Life Sciences Education* 11 (2): 113–29. https://www.lifescied.org/doi/10.1187/cbe.12-03-0031.

Lee, D. H. L., and I. Shaari. 2012. "Professional Identity or Best Practices?—An Exploration of the Synergies between Professional Learning Communities and Communities of Practices." *Creative Education* 3 (4): 457–60. http://dx.doi.org/10.4236/ce.2012.34070.

Maier, M., K. R. Rowell, and R. H. Macdonald. 2021. "Disciplinary Societies and Community Colleges: Partners in Progress." *New Directions for Community Colleges* 2021 (194): 159–71. https://doi.org/10.1002/cc.20461.

NRC (National Research Council). 2012. *A Framework for K–12 Science Education: Practices, Crosscutting Concepts, and Core Ideas*. The National Academies Press. https://doi.org/10.17226/13165.

Reinholz, D. L., and N. Apkarian. 2018. "Four Frames for Systemic Change in STEM Departments." *International Journal of STEM Education* 5, article 3: 1–10. https://doi.org/10.1186/s40594-018-0103-x.

Reinholz, D. L., and T. Andrews. 2024. "Change as a Scientific Enterprise: Practical Suggestions about Using Change Theory." *Cell Biology Education–Life Sciences Education* 23(1): es1. https://doi.org/10.1187/cbe.23-06-0103.

Reinholz, D. L., I. White, and T. Andrews. 2021. "Change Theory in STEM Higher Education: A Systematic Review." *International Journal of STEM Education* 8 (1): 37. https://doi.org/10.1186/s40594-021-00291-2.

Wenger, E., R. McDermott, and W. M. Snyder. 2002. *Cultivating Communities of Practice: A Guide to Managing Knowledge*. Harvard Business School Press.

Part Three

Driving Cultural Change through Deliberate Interorganizational Collaboration for Inclusive Environments

Introduction to Part III

Achieving transformative change in STEM inequities requires fostering interorganizational collaboration and promoting professional development to tackle the persistent underrepresentation of women and people of color —especially women of color—in education, careers, and leadership roles. Addressing these inequities requires systemic culture change, and STEM professional societies are uniquely positioned to lead such efforts due to their authority in setting disciplinary standards and their influence across sectors. This part examines two case studies that highlight strategies for advancing equity and fostering collaboration in STEM. The first explores the Alliance to Catalyze Change for Equity in STEM Success (ACCESS) and analyzes its multi-organizational collaboration through four organizational frames—symbolic, structural, human resources, and political—to identify key lessons for sustaining equitable partnerships. The second focuses on the ACCESS+ initiative—Amplifying the Alliance to Catalyze Change for Equity in STEM Success. The ACCESS+ initiative combines critical theoretical frameworks, tools for reflection and assessment, and a community-based approach to empower change-makers in thirty-three STEM professional societies. Together, these case studies provide actionable insights for building inclusive networks and fostering collaborative efforts to advance STEM reform.

11

The Alliance for Catalyzing Change for Equity in STEM Success (ACCESS): Lessons Learned and Implications for Organizational Collaboration

Lucas B. Hill and Gretalyn Leibnitz

The Alliance to Catalyze Change for Equity in STEM Success (ACCESS) was funded by the NSF (NSF 1744098) to "create a community of practice of diversity leaders within scientific societies to collect, discuss, and share tools and resources that would help them address the needs and concerns of those historically excluded within the scientific community." An organizational analysis was commissioned in 2021 to elucidate a common understanding of the challenges faced and benefits experienced by ACCESS leaders and to provide recommendations for future directions. In addition to giving ACCESS leaders insight into how the alliance functioned and potential pathways forward, we believe the resulting analysis provides a window into a successful collaborative structure that could be useful to other leaders of collaborative initiatives.

In this chapter, we seek to provide insights into starting and scaling a multi-organizational collaboration. We use Bolman and Deal's (2017) four frames to examine the ACCESS organization, a network of biological professional and disciplinary societies seeking to improve DEI within their organizations. By sharing this work, we hope to provide insight for professional society and other higher education leaders involved in creating and sustaining collaborative relationships for advancing collective goals. Our guiding questions for this chapter were twofold: (1) What organizational features should collaborative change projects be aware of while advancing collective change? (2) What are the implications of the lessons learned from the ACCESS case?

Background of the Study

As a means of improving America's global competitive advantage through a diverse and well-trained STEM workforce, there is persistent pressure to improve STEM education (Atkinson 2012; NAS et al. 2007; NCSES 2023). Many early efforts to reform STEM education consisted of focused, institution-based interventions that were intended to be shared at a wider scale across institutions (Kezar 2011); yet widespread change has not occurred as expected (López et al. 2022). More recently, scholars and practitioners have called for more systems-oriented and collaborative change initiatives that span institutions and even sectors to more effectively promote systemic change (NASEM 2016; Robinson 2022). It has been reasoned that multi-organizational collaborations foster more efficient reform through (a) the sharing of knowledge, expertise, and resources; (b) fostering innovation and better problem-solving capacity; (c) building greater resilience in times of turbulence through social connections and enhanced organizational agility; and (d) ultimately, creating capacity for large-scale, systemic, social change (Agranoff 2012; Bryson et al. 2015; Foster-Fishman et al. 2001). Through ADVANCE, Alliances for Graduate Education and the Professoriate, Inclusion across the Nation of Communities of Learners of Underrepresented Discoverers in Engineering and Science (INCLUDES), and Research Coordination Network (RCN) funding vehicles, the National Science Foundation has launched numerous programs that center collaborative efforts in reform design. However, despite the ensuing prevalence of collaborative change in STEM and emergent research (e.g., Gehrke and Kezar 2016; Hill 2020; Kezar 2014), there is still limited knowledge about how context and function influence collaborative entities. To better understand and effectively leverage the power of collaborations, the context and functioning of successful collaborations need to be further explored.

Case Study Narrative

ACCESS was originally composed of six organizations concentrated in the biological sciences: American Society of Biochemistry and Molecular Biology (ASBMB), American Society of Cell Biology (ASCB), American Society for Pharmacology and Experimental Therapeutics (ASPET), Biophysical Society (BPS), The Endocrine Society (TES), and the Scientific Careers Research and Development Group (SCRDG) at Northwestern University. ACCESS members (approximately 20 individuals at the time of the study) consisted of a mix of higher education faculty and professional society staff, both of which often served as leaders of minority-action committees (MACs) within their societies. A major portion of the project budget was originally intended for three in-person convenings, with minimal funding associated with paid personnel time. However, project needs and evolution prompted subsequent budget reallocation.

Members convened once in-person, in collaboration with the HHMI, to create committees to work on different aspects of the grant project. However, due to complications raised by Covid-19, the group switched to online convenings and later ended the multiple committee format, instead opting for a core community of practice that met regularly online. During these calls, members effectively shared knowledge and worked on collective products such as shared measures and publications (e.g., Segarra et al. 2020ab).

By the conclusion of the grant in 2021, ACCESS had created a database of best practices delivered via multiple communication channels (e.g., website, talks, presentations, and publications) on topics such as travel awards and society speaker diversity. ACCESS also provided guidelines and evaluation metrics to examine the broadening participation activities of professional societies and demonstrated the utility of an alliance comprised of STEM societies. Overall, ACCESS resulted in a strong community of committed and collaborative individuals who were invested in advancing the careers of scientists from historically excluded groups within their organizations. As of 2024, ACCESS has evolved into a new NSF-funded project (NSF 2017953) called ACCESS+, or Amplifying the Alliance to Catalyze Change for Equity in STEM Success. ACCESS+ has grown to over thirty societies, including disciplines outside of the biological sciences. (For additional information on the evolution of ACCESS, see Chapter 12 on ACCESS+.) The focus of this chapter is information derived from the organizational analysis of the original ACCESS project.

Organizational Analysis

We used the four frames of Bolman and Deal (2017) to structure our organizational analysis of ACCESS. Bolman and Deal's (2017) four-frame model has proven useful in the STEM reform context (Reinholz and Apkarian 2018; Reinholz et al. 2021). Through reflection on the symbolic, structural, human resources, and political frames, this model helps make explicit the main currents that operate within organizations. The *symbolic frame* consists of the grander purpose of an organization, its origin story or organizational saga, its culture, and the ethos that guides its operation. The *structural frame* represents the nuts and bolts of the organization, including the main organizing components, its mission and goals, and the guiding policies and procedures. The *human resource* frame focuses on the people of an organization, their needs, and the motivations that guide their engagement in the organization. Last, the *political frame* deals with the power dynamics, resource constraints, and conflict that appear in organizations and affect operations.

Our organizational analysis involved three steps. First, we conducted a thorough review of core ACCESS documents, artifacts, and publications, paying close attention to key variables and themes related to organizational dimensions. This step allowed us to identify the most salient aspects of ACCESS and notable themes to incorporate into the following step. Second, we had detailed conversations with ACCESS leadership ($n = 6$; five faculty members and one professional society staff) to explore the history of ACCESS, current and original goals, progress toward goal achievement, organizational dynamics, and the vision for future organizational structure and operations. These conversations lasted between 30 and 60 minutes, were recorded, and transcribed verbatim. We used a thematic analysis approach to identify core themes and findings. Last, using documents and interview data, we conducted and compiled a thorough analysis of the current organizational structures and collaborative dynamics of ACCESS. This included the mapping of ACCESS members and key personnel, including roles and responsibilities; using the symbolic, structural, human resource, and political frames to organize and interpret emergent findings and themes; and conferring with ACCESS leaders to check the accuracy of our findings.

Below, we report the findings of our organizational analysis of ACCESS using the four organizational frames. We conclude the chapter with key recommendations for those engaged in higher education collaborative reform.

Findings

Symbolic Frame

As a reminder of the definition provided in the introduction, the symbolic frame consists of the grander purpose of an organization, its origin story or organizational saga, its culture, and the ethos guiding its operation.

Minority Action Committee (MAC) Roots

ACCESS was deeply rooted in prior efforts to create MACs in professional societies starting in the 1980s and efforts to create a super MAC (i.e., a group of MACs from different societies) in the 1990s and early 2000s. In the words of one interviewee, "At our society we have a grant called Super MAC that emerged from veteran society members who believed that we can do this work (i.e., address DEI issues) in a more effective way, pooling our resources to influence or impact inclusion in our societies. I think that is where we came from."

The underlying goal of these past efforts was to advance collective change to the benefit of minority scientists, an ethos that undergirds ACCESS's mission. By observing ACCESS meetings, it was readily apparent that society representatives in ACCESS and ACCESS leaders strongly held to its historical roots and likewise prioritized the advancement of scientists from historically excluded groups. Thus, the culture of ACCESS was one of commitment to improving opportunities and STEM culture for scientists from historically excluded groups.

With that said, one interview participant mentioned that there was some skepticism among those connected to ACCESS "that something like this could be put together," based upon somewhat ineffective, earlier attempts at building super MACs. However, interview participants indicated that prior efforts provided the fertile ground on which to continue, not replace, previous goals and related work.

Birth of an Idea

Directly stemming from prior work within and between MACs, it was important to capture ACCESS's initial vision as described in the grant proposal as a "community of practice of people working together with a common goal" to share best practices and support the work of DEI advancement within professional societies. This focus on the development of ACCESS was commonly held and articulated by most interviewees. Most participants echoed the words of one interviewee:

> Although many societies were working on minority (DEI) issues, we were doing so independently, not really using best practices (maybe not even knowing what practices were evidence-based), not even talking to each other on what we were doing, how we were doing things, and whether the things we were doing were working; and unless we did something different, we would just keep doing the same things. Working together was a way to crowdsource information, resources, and practices.

This community vision was initially advanced by the principal investigator (with the support of other key stakeholders within the ACCESS initiative) and constitutes a major aspect of its operating ethos.

Minority Scientists

An important theme related to the symbolic frame is how ACCESS has served as a refuge and safety net for faculty from historically excluded groups. In short, many of those engaging in ACCESS were faculty from historically excluded groups themselves, which influenced its culture and priorities and integrated the shared experiences of ACCESS members into advancing DEI efforts within and between professional societies. One interviewee noted, "this work taps into my heartstrings," and another stated, "I have learned a lot more about the scholarship in the field, thought more about the work, and it has given me better language to talk about things that I knew were issues but didn't quite know what to say about them. I think that is true for everyone." This lived experience was both explicit and implicit within ACCESS and was the lens by which other organizational components were viewed and engaged.

STEM Disciplines and Faculty Ethos

Another important organizational cultural element embedded within ACCESS was a STEM disciplinary and faculty perspective, one that focused on data-driven arguments, publications, conference presentations, and the pressures of busy schedules that lack the boundedness of typical nine-to-five positions. Even though many organizational representatives in ACCESS worked for their respective society (i.e., individuals who were neither faculty nor conducting research), the focus was on the STEM research-faculty lens, which helped define the priorities of ACCESS. Although each society brought a unique disciplinary perspective to the table, all societies were concentrated in the biological sciences and focused on a research STEM faculty perspective (i.e., the importance of evidence, rigor, and career advancement through grants and publications) that enforced the other organizational frames described below and, at times, was not perfectly aligned with MAC roots and DEI priorities.

Professional Societies

Each professional society had a unique mix of organizational creation stories, characteristics, and operational norms. ACCESS was the amalgamation of the symbols and culture embedded within multiple organizations represented. Thus, ACCESS was a combination of shared (i.e., MACs, super MACs) and individualized (personal and organizational specific) symbols that were woven together as it evolved.

Summary

Symbolically, the grand purpose of ACCESS was to promote opportunities for scientists from historically excluded groups and foster an inclusive STEM culture. Society representatives sought to achieve this shared goal through a community of practice committed to sharing and applying evidence-based practices. The motivation for this work was anchored in the lived experiences of the leaders of ACCESS, many of whom were STEM faculty from historically excluded groups, which meant that the application of data and ultimately the dissemination of outcomes through publications and conference presentations were salient to their careers. Critical to this symbolic frame was the successful weaving together of individual symbols (i.e., personal and organization specific) in a functional collaborative gestalt. The major takeaway for those leading collaborative change is the necessity to unpack organizational and personal symbols to guide structures and operations within the other organizational frames described below.

Structural Frame

As stated above, the structural frame represents the nuts and bolts of the organization, including the main organizing components, its mission and goals, and guiding policies and procedures.

Collaborative Structure

A small number of individuals represented each member organization and attended meetings (i.e., advisory board, steering committee, and online full group gatherings). At the time of the organizational analysis, there were about twenty individuals active in ACCESS. The advisory board was composed of seven individuals who provided feedback about ACCESS's mission, goals, and activities. The primarily volunteer Steering Committee constituted the core group of ACCESS who met more regularly and advanced its mission. ACCESS was formally led by the PI, but due to health issues, the primary operational leader for most of the grant was a co-PI. There were two alternative PIs at the primary grant institution over the five years.

Resource Delimiter

As valuable and vital as the ACCESS grant proposal was, it also served as a basis for a major limitation to organizational infrastructure. The proposal budget included three ACCESS-leader convenings but did not provide fiscal support needed to facilitate the work done between convenings; namely, funding was absent for the essential resources needed to build the complex, collaborative infrastructure (e.g., administrative support, faculty time buy-out, marketing and communication expertise). That said, additional pressures came to bear on the project in the form of Covid-19. The pandemic required that the ACCESS leaders pivot two of the in-person convenings online, which required fewer fiscal resources, made it easier for members to participate, and even extended the funding period. Yet, the lack of core resources to support holistic organizational development had ongoing implications for the ACCESS's structure.

Enacting ACCESS's Mission through Committees

Early in ACCESS the intent was to create and run committees to advance concrete actions related to ACCESS's mission. In 2018, ACCESS worked with HHMI to host a convening of professional societies, which included the seed conversations and connections for several action-oriented committees including the Dissemination Committee, Evaluation Metrics Database Committee, New Partner Committee, and Sustainability Committee. These committees were in addition to the Advisory Board and Steering Committee. However, in the words of one interviewee, "The momentum died when people got siloed into committees. It just didn't work." Despite initial traction in some of the committees, interviewees indicated that ACCESS members were spread too thin, and the committees mostly dissolved into a single steering committee that consisted of those who regularly attended meetings. One participant explained:

> I think the only way for us to rally and drive the progress forward efficiently was to downsize, there's only so much that you can do with too many people and too few heads leading ... because then the person who's leading then has to direct the people that are in the committees plus their own work. And so at some point, there's just an efficiency equation there where it's just easier for someone to do the work, you know, with a smaller group of people.

In addition, one interviewee mentioned that they "didn't think the working groups ... could actually work ... [because] in many respects [they were] premature from the standpoint of what the committees were being asked to do ... because you had very different visions of what the role or what the purpose of ACCESS was." Yet, another participant stated, "I think that I think the mission has stayed pretty consistent ... I think the thing that's evolved over time is ... the objectives [and actions] that go under that." Thus, the first few years of ACCESS, like many organizations, were marked by its membership trying to determine what to focus on with the resources available under the umbrella of an overarching goal.

While the intent of this section is not to evaluate the success of ACCESS in accomplishing its goals, it is important from a structural perspective to note that despite the success of the committee model in academic settings, this model was unsuccessful as an organizing structure. The ineffectiveness of the committee structure was due to a lack of agreement about the purpose of ACCESS. The Steering Committee, which had served as a community of practice and as an organizing force to accomplish tasks (e.g., the travel award program and publications), proved more effective.

Support Resources

One of the major reasons (if not the major reason) for the lack of success with multiple committees was the limited availability of time and capacity from society representatives and faculty members. At the time of the grant proposal, some believed that society representatives would have some of their formal work responsibilities directed to ACCESS engagement. However, societies were already stretched thin in terms of paid time to cover their many responsibilities. As one interviewee mentioned, "[society staff] are constantly asked to do a little more with a little less ... [the] amount of time people have to sort of act [on] these external collaborations, I'm sure is smaller and smaller ... wedging this into an ever-dwindling amount of time."

Similarly, interviewees mentioned that "we underestimated the amount of effort and time and funds that [the proposed work] would take," "we didn't really know what was involved or how much work was going to be [involved], " and "what we were able to achieve was determined by the realities of having a multi-organizational project with only a few people deeply involved." Several interviewees also mentioned staff turnover, which resulted in difficulties integrating society staff members into the evolving initiative. Furthermore, interviewees expressed concerns that the proposed activities in ACCESS were not always aligned with the expertise, self-efficacy, and capacity of its membership (e.g., building a database). For instance, one interviewee mentioned that many of the early meetings had a lot of sustained silence due to a lack of understanding and consensus about its purpose and direction. Overall, ACCESS had to learn both how to engage its members toward common goals and how to do so within realistic and limited capacity.

ACCESS was able to secure some logistic support early in the project and at the time of analysis had a few undergraduates in an assistance role, but largely, there was not sufficient logistical support to run a multi-organizational structure. In addition, both the PI and co-PI institutions had logistic support roles in the project but, at times, logistic support roles were inefficient or not aligned, which was succinctly described by one interviewee as "administrative red tape." These complications were further exacerbated due to leadership fluctuations throughout the project. Thus, structurally speaking, there were complications with institutional administrative support and a misalignment of effort allocation in the context of ACCESS goals where most involved were participating in a voluntary capacity (i.e., without compensation), fueled by their interest and passion for broadening participation in STEM. Due to leadership transitions a major brunt of the organizational operations of ACCESS fell to the co-PI, who had minimal paid time on the project. In short, not having sufficient core operations support had structural implications for the organization.

Coinciding with the lack of support resources was the lack of formal connection between the work of the volunteer MAC leaders in ACCESS with formal professional society leadership. Although all interviewees were pleased with the outcomes of ACCESS and felt that the project had met its goals for the most part, interviewees mentioned that they felt it struggled because of a lack of formal connection to professional society leadership. When ACCESS interviewees were asked what type of formal connections they thought would have been helpful, hindsight responses included: (a) inviting society leaders to ACCESS events, (b) providing quarterly reports to society leaders from ACCESS (not the MAC leaders), (c) inviting feedback from the society leaders to ACCESS leaders, (d) having the MAC chairs report to the society leaders about the benefits of engagement in ACCESS, and (e) specifically addressing the value of the professional society engagement in ACCESS with society leaders—especially in a resource competitive environment—so that time, information, and resources could be prioritized for ACCESS work. That said, it was clear that when formal leadership was aware of MAC efforts, there was more support and more investment in the DEI work. One interviewee mentioned that once their society council received the ACCESS travel award paper, the society council determined a need to streamline the travel award process and update language (e.g., remove "minority" from award labels to not minimize these awards and the people who received them). Support, such as that provided by this society council, created a positive feedback loop that resulted in greater interest in and attention to DEI work. Thus, formalizing the alignment between ACCESS and professional society leaders would have resulted in more synergistic DEI work, greater awareness of ACCESS efforts, and broader impacts.

Summary

Structurally, a lack of core resources limited the development of ACCESS. The roughly twenty-person ACCESS team was led primarily by a long-term operational co-PI and a more traditional committee structure that was largely unsuccessful. The Steering Committee core, as informed by an advisory board, provided the infrastructure necessary to accomplish the work of ACCESS. The proposed general ACCESS mission evolved to reflect concrete actions influenced by the priorities of, and accomplished in, the volunteer time available to passionate academic faculty leaders and their students. A greater ACCESS impact could have been achieved with a stronger connection between ACCESS leaders and professional society leadership. So, in retrospect, ACCESS's accomplishments were less the result of adequate planned infrastructure and fiscal support and more a result of the effort and passionate commitment of its volunteer faculty leaders.

Human Resources Frame

As indicated above, the human resource frame focuses on the people of an organization, their needs, and motivations guiding their engagement in the organization.

Motivation

For those involved in ACCESS, motivation was by far the strongest organizational feature. As already mentioned, most ACCESS members spent their own free time participating and advancing the mission. A common theme with interviewees was expressed by this participant, "The work of ACCESS has been pretty central to what I did." It was clear by observing ACCESS meetings and from interviews that there was a deep and sustained passion for improving DEI within STEM and professional and disciplinary societies. ACCESS provided members, many who were early career professionals, with "an incredible sense of community." In the telling words of this interviewee, "Everyone who has been involved has learned a lot and offered each other a community of like-minded people. Now I know that I am not a weirdo swimming against the stream of STEM—there are a lot of people that want STEM to change and that has been a positive effect for me, and I can't envision otherwise for anyone else."

ACCESS members "produce things, build on ideas, they're creative, they get things done … that has been a huge synergy." Similarly, another participant commented, "Because of the similar work that we're doing, that drew us to get closer and I think that's just really powerful. And that type of synergy, it really drives my desire to work harder, and do more, so that's been amazing."

Generally, interviewee participants commented on the fact that ACCESS engagement is "helpful for me to see what the other societies are doing and to be able to visualize where we are," since being in ACCESS "[gives] a better sense of where we all fit in the bigger scheme of things."

However, there were some concerns mentioned by interviewees about the volunteer nature of ACCESS and the fact that societies have differences as well as similarities. One participant noted that " … eventually, you will burn out the volunteer energy … unless there is some return on investment." Another interviewee commented:

> Our members can look really different, depending on, you know, what we're talking about, and where they're positioned. And I think that's a good thing to keep in mind, too, because we are hoping that what will come out of this can be at least to some extent, adapted or made generalizable to different societies in different groups. But understanding that our memberships are not a homogeneous pool of people, I think that has been an important lesson.

Thus, there were benefits to having highly motivated members but also the limitations of volunteerism embedded within considerable variability in the society space.

Organizational Identity

The distinction between individual motivators and organizational partner motivators (i.e., the professional societies) was not entirely clear. As one interviewee put it, "[participation in ACCESS] is almost completely at the individual level," meaning that societies had less stake as organizations versus individual society representatives. For most of ACCESS's history, members were best described as boundary spanners who connected their society or institution to ACCESS rather than formal organizational representatives. These boundary spanners wore different hats, with some holding staff positions at professional societies and others being faculty who held temporary leadership roles (current or prior) within their professional society or had a passion for advancing DEI within their field. This may appear somewhat trivial, but it had implications for how ACCESS envisioned itself as an organization and where membership resided, either individual or organizational. In addition, it had consequences for how the benefits of ACCESS participation were perceived, either as individual benefits that could also benefit home organizations or as organizational benefits that were translated through society representatives.

Prior Connections

Another important factor that informed interactions within ACCESS was the prior connections that existed before the formation of ACCESS. Some of the societies had pre-existing relationships with other societies and individuals in ACCESS, which shaped the initial formation and society invitations. This prior history initially helped provide some footing for ACCESS to function, although prior history was not universal across its membership.

Interviewees also noted that it was challenging to engage new people in ACCESS due to turnover. As indicated by one interviewee:

> One of the bad things of ACCESS is that there is a lot of turnover in some of the societies, and the membership, and that has a cost associated with it because whenever there is someone new in ACCESS it takes a while to get to know everyone and whenever there is someone new, they are not quite sure what to do; however, all the members are very welcoming. It is not a scary place.

Although networking was a recognized highlight of ACCESS, "just throwing a new person into the system is not the best way. We need to work to onboard people better." So, limited strategic connectivity, prior network connections, and a lack of process for building new connections served as a delimiter that required activation energy to build additional cross-network ties.

ACCESS Roles

Six primary roles of ACCESS members were identified: leader (formal and informal), convener, contributor, boundary spanner, peer support, and administrative support. First—based on interview data, documents, and observation—it was clear that there were several within ACCESS who had both formal (i.e., PI, co-PI, committee chairs, manuscript leads) and informal *leadership* roles (i.e., thought leaders). Second, some in ACCESS took on a *convening* role by bringing together the whole membership or parts of it to advance its mission and goals. This role often overlapped with leadership responsibilities. Third, some held the role of *contributor*, whether it was meeting participation, manuscript advancement, or other responsibilities. The basic idea was that many in ACCESS were active contributors through regular and sustained activity. Fourth, most within ACCESS also served a *boundary-spanning* role by formally or informally connecting their society or institution to ACCESS, both in bringing things from their home institution into the ACCESS space and things from ACCESS back to their organization or institution. For example:

> One of the benefits of ACCESS is that we do not have dedicated funding for publications at my university, so to have the grant pay for publication costs has been critical. So, again, I am advocating university-wide, that this needs to be covered. The provost office needs to know that for us in our discipline it is critical that we get that money for publication expenses.

Another responded that "because I know how the other societies in ACCESS are handling things I have been able to speak in faculty affairs, and other places, with knowledge about how other societies are doing things, and that is good, so that I can try to request resources for faculty."

It was not uncommon for interviewees to comment on how their ACCESS work served them in other professional roles of their life, whether as a faculty member or in specialized academic institutional leadership roles (e.g., "I have been able to leverage the work of ACCESS at my home institution").

Fifth, as the community in ACCESS evolved and expanded there were several who excelled at providing *peer support* through career advice, a listening ear, and helping to troubleshoot individual and organizational challenges—both internal to the project and other areas of STEM professional life. Last, some in ACCESS provided the role of *administrative* support such as coordinating meetings, taking care of grant-related paperwork, and managing the budget. In many ways, the six primary roles worked in tandem to advance the mission of ACCESS. However, there was considerable tension at times between the various roles because of differing visions for the work and mission of ACCESS, leadership incongruities, and layers of bureaucratic processes within participating institutions.

Group Norms

It was clear from observing ACCESS meetings that members were respectful to each other and, as mentioned above, displayed congruous beliefs in the mission of ACCESS and DEI in general within STEM disciplines. In particular, the co-PI effectively facilitated meetings to encourage inclusive participation and that voices were heard from all present. As one interviewee explained, "I think they've done a great job of really engaging everybody ... they're a community that engages and welcomes everybody's thoughts." Another interviewee concurred, "As far as I have seen, we are all very open in ACCESS; we listen to what others are doing in other societies." Another cited that there is evidence of effective collaboration and knowledge exchange "because you can already see the cross-fertilization, the ideas that people are sharing back and forth that they'd never have thought to have access."

However, as mentioned above, group norms (i.e., respect, inclusive participation, active listening, knowledge exchange) developed over time as more and more participants became comfortable engaging in the resultant community of practice. This was at least partly due to the reduction of working groups into a single steering committee that supported the development of group norms and increased collegiality. In addition, interviews made it apparent that, like many organizations, there are public group norms and behind-the-scenes group norms among smaller pockets of ACCESS members. The inference here was not that subgroups have some innate nefarious or dubious purposes but that there were interlocking layers of large and small group norms at play within ACCESS. For example, smaller groups formed to provide each other support, often because of their experiences as marginalized individuals within their organizational or institutional roles. Other subgroups formed for specific logistic or other purposes. However, the group norms that emerged were often a function of the task at hand. One interviewee put it this way, "I know there was formal structure in the beginning—especially when we were setting up the database—but that has not continued. I think of it as ACCESS together working on a particular task. ACCESS is not hierarchical. We are all invited to and are engaged in the work of ACCESS."

Last, conflict is inevitable in collaborative work, especially since reform initiatives often overly focus on programs and change levers and pay insufficient attention to establishing effective collaborative dynamics during initial development stages. ACCESS, like many DEI projects, experienced conflict in the first years of the project. In the words of one interviewee, the early years of ACCESS were "a mess," especially within project leadership. We unpack the ACCESS early years organizational mess in the political frame that follows.

Summary

Not surprisingly, given the complex nature of human beings, the human resource frame contained multiple elements that helped to frame the form and function of ACCESS. ACCESS was built by individuals with strong motivations to improve DEI dynamics within their societies while also reflecting diverse organizational identities, connections (current and prior), and roles. This resulted in layers of group identity and norms, albeit housed within an eventual holistic culture of trust and respect. Leaders of collaborative change initiatives would be wise to mindfully map out the people involved in the reform project, including their prior history, identities based on different organizational forms and demographic characteristics, and desired and realized initiative roles. In addition, in alignment with the symbolic frame, it is critical to explicitly state and collectively define group norms that are congruent with the purposes of the change initiative. Leaders must also map and strategically plan time allocations across the collaborative to make the project a success, knowing that it will be a mixture of funded and unfunded effort and require (and where possible) aligning initiative and regular job responsibilities synergistically to account for the time constraints of individuals. Overall, the human resource frame reminds change leaders of the importance of centering people in change work and that just getting everyone in the room is not the conclusion but the start of collaborative work.

Political Frame

As a reminder, the political frame deals with power dynamics, resource constraints, and conflict that appear in organizations and affect operations.

Power Dynamics

It is important to acknowledge that ACCESS members were at multiple stages of their careers. The career stage difference between the PI (established, late career) and co-PI (early career) likely had the greatest impact on the organizational structure and operations of ACCESS. At multiple times in the project, the PI had to take medical leave, which caused unpredictability for ACCESS members and leadership challenges. As one interviewee explained, the first few years of the project "was just downright bumpy." Another interview participant further explained the impact of the leadership transitions this way:

> There was a period of time where, at least in my memory, it was like, okay, the PI is out, so we're going to stop for a little bit. Now the PI is back, so let's try to hurry up and do some things. And then the PI is out again, so let's stop. Once we stopped kind of trying to adapt in that way, but just let the co-PI kind of lead things, I think that made it smoother, because I think it was the unpredictability, and people just kind of not being able to wrap their heads around what was coming next that made it more difficult.

In short, it was difficult for the project to go through multiple transitions where the more experienced, tenured PI—who had a clear vision for the project but was not always present—left the early career faculty member and co-PI to take the reins within a hierarchical leadership structure. For example, this leadership dynamic played out during the creation of the website. Specifically, leadership disagreement resulted in a lack of consensus about the website's purpose in advancing the goals and objectives of ACCESS, how and when to engage and direct paid consultants to build the website, and what key features it should possess. This created considerable stress for a smaller group of ACCESS members working on the website. As one participant put it:

> That was almost fatal … engaging [the website developers] and then disengaging them, that came really close to several [saying] this ain't gonna work, I'm out of here. Yeah, it was really bad. You know, I think most of the general membership was shielded from it. But there are a few of us that were immersed in it. It was not pretty … that almost took [ACCESS] down.

The back and forth, while eventually resolved, was a major impediment for ACCESS and resulted, at least temporarily, in a polarization of leadership that left personal distress and distrust in its wake. ACCESS, as supported by a grant designed to mainly host in-person convenings, was unable to spend sufficient time planning for disagreement and building a leadership structure that would have clear and consistent roles and responsibilities with contingency plans for leadership transitions.

Scarce Resources

While resource constraints have been discussed above, it is important to note the political ramifications of scarcity within ACCESS. The very essence of the political frame is the availability and distribution of scarce resources whereby individuals within an organization seek to obtain and apply resources toward individualistic and group goals. This dimension has been particularly manifest within budgets at the PI and co-PI institutions, especially as the co-PI was left on multiple occasions to run the day-to-day operations of the organization without direct fiscal oversight (even if temporary). Thus, resource allocation became a significant sticking point and source of tension surrounding ACCESS operations.

Behind the Scenes

As noted above, subgroups formed both structurally (committees) and organically (collegial support). Like most organizations, there were two layers within ACCESS, one that consists of the entire organization in the nearly public domain (i.e., large world) and the smaller pockets of interaction and engagement (i.e., small world). As previously mentioned, there were collegial interactions within the large-scale world of ACCESS. However, there have also been, at least initially, complicated interactions in the small-world domain. Yet, despite the occurrence of organizational stress, these smaller negative interactions prompted small-world support from pockets of members, which encouraged tighter bonds between members to the benefit of the larger organization. In addition, it is important to note that there were other types of small-world interactions related to the administrative and logistic aspects of ACCESS, such as coordinating reports, discussions about funding, and so forth that have had varying levels of conflict and resolution. Small-world engagement is not only a functional means of social support but also a critical factor in helping execute the work of the organization.

Summary

ACCESS members existed within large- (public) and small-world (private) layers of the organization, which required the navigation of power dynamics and scarce resources. The power of ACCESS was not found in the convening of the entire organization alone but by finding ways for smaller circles to develop relationships and increase investment in organizational productivity. With that said, ACCESS committees were intended to develop small-world connections but were largely unsuccessful. Yet, the Steering Committee developed into a stable community of practice that consisted of caring and supportive colleagues. Furthermore, ACCESS members encountered conflict concerning both resources and role-based hierarchy that was eventually resolved but still had a large impact. Although conflict is an inevitable part of collaboration, not many project leaders plan for disagreement or develop transparent strategies for navigating differences or the equitable distribution of resources. Often the focus is prematurely on completing deliverables without the proper preparation of people (i.e., change management) for the work proposed (i.e., project management). Spending more time building the interpersonal infrastructure of trust, empathy, and understanding of each other's life contexts—while working to concretely define project deliverables and align roles with strengths, skills, and resources—goes a long way toward reducing and minimizing the impact of conflict.

Discussion and Reflection Questions

The following questions are designed for the reader to explore the organizational dimensions of their collaborative projects and initiatives. To gain the most benefit, we recommend that the reader circulate these questions with their teams to discuss perspectives and then examine the implications for further and continual organizational development toward change goals and priorities.

Symbolic Frame

- What are the historical roots of your collaborative change projects (new or emergent)?
- To what extent is there a shared vision for desired goals and outcomes?
- How, and to whom, is the work considered important and meaningful?

- Where might there be common traditions, ceremonies, and rituals, and how might these be leveraged for change?

Structural Frame

- What is the structure, including leadership, of your collaborative change initiative?
- What are the collaborative policies, procedures, and practices?
- What resources are necessary to accomplish change goals and how are they distributed?
- What are the boundaries of the collaborative?

Human Resource Frame

- What motivates the individuals in your collaborative initiative?
- What are the motivations of your collaborators (current and potential)?
- What are your foundational group norms?
- How were your collaborators previously connected?
- How will you make decisions about filling roles deemed critical for the collaboration?

Political Frame

- How will you examine and manage power dynamics and their impact on the collaborative change initiative?
- How will sufficient resources be secured and allocated equitably?
- How will the development of interpersonal relationships and investments facilitate the work of the collaboration and lead to productivity?
- How will you address divergent interpersonal dynamics and conflict?

Conclusions

We conclude this chapter with recommendations and lessons learned gained from our organizational analysis. We used Bolman and Deal's (2017) four frames to examine the organizational components at play that influence what is chosen as the focus for the work, how the work is accomplished, and the resultant impacts.

Symbolic

- *Pay attention to prior organizational history and connections* to maximize preexisting foundations in service to your change goals and work to address the potential lack of integration across constituents.
- *Collectively define and record the collaborative's creation story* such as what motivated its creation and the heart and soul that makes it unique in the STEM reform community.
- *Co-create and elevate an equitable organizational culture* that pays attention to disciplinary differences and constituent perspectives as well as establishes normative behaviors.

Structural

- *Build policies and practices to integrate and onboard new members and partners* into the organization to reduce orientation time and to ensure the alignment and buy-in of key norms and behaviors.
- *Explore potential organizational formats* that can support the change work after external funding ends, such as a membership dues model, fee-for-service, and the integration into an existing host organization that can help bring in resources and provide logistic administrative support.
- *Develop a sustainability plan* that identifies ongoing fiscal support to provide sufficient and consistent logistic support to run project operations. We strongly suggest that initiatives explore how core staff can be hired to run operations and free up time for members and partners to focus on goal achievement.

Human Resources

- *Anticipate and plan for conflict* by spending critical time in advance building trust and empathy as well as aligning personal strengths with project and organizational time and resources. It will also be critical for initiatives to have a clearly defined and agreed-upon process by which to address and resolve conflict as well as make decisions about its mission, strategic actions, and administrative and logistic support.

- *Create a shared leadership model* that both distributes the workload and advances DEI components within your collaborative change project. We strongly recommend that this leadership structure has a plan for leader, constituent, and partner turnover as well as instances of unexpected non-participation to prevent pauses of the work.
- *Align and prioritize the work of the collaborative with the professional goals and activities of the people involved* so that change activities synergize with their primary work responsibilities and the work of the collaborative is not overly reliant on volunteer labor. Part of this will require that you map and articulate the benefits of membership and engagement so that current and potential collaborative members can more easily justify their involvement.

Political

- *Come to a consensus about the boundaries of the collaborative change organization*, including leadership structure and membership, since funded projects will often sprout new related initiatives that could create a messy political landscape. This includes determining how each of the related initiatives and organizations will interact and evolve over time.
- *Transparently examine power dynamics within the organization,* including a review of role-, demographic-, and expertise-based characteristics and the resultant configurations of power distributions.
- *Provide ongoing support for small and large worlds.* As your change initiative grows and matures, it will be important to pay attention to holistic organizational dynamics as they shift and evolve. In addition, it is vital to provide space for smaller pockets of individuals to connect and develop meaningful relationships as they seek to address sub-components of the collaborative's mission and to regularly connect these smaller groups to the greater whole.

References

Agranoff, R. 2012. *Collaborating to Manage: A Primer for the Public Sector*. Georgetown University Press.

Atkinson, R. D. 2012. "Why the Current Education Reform Strategy Won't Work." *Issues in Science and Technology* 28 (3): 29–36.

Bolman, L. G., and T. E. Deal. 2017. *Reframing Organizations: Artistry, Choice, and Leadership*. John Wiley and Sons.

Bryson, J. M., B. C. Crosby, and M. M. Stone. 2015. "Designing and Implementing Cross-Sector Collaborations: Needed and Challenging." *Public Administration Review* 75 (5): 647–63. https://doi.org/10.1111/puar.12432.

Foster-Fishman, P. G., S. L. Berkowitz, D. W. Lounsbury, S. Jacobson, and N. A. Allen. 2001. "Building Collaborative Capacity in Community Coalitions: A Review and Integrative Framework." *American Journal of Community Psychology* 29 (2): 241–61. https://doi.org/10.1023/A:1010378613583.

Gehrke, S., and A. Kezar. 2016. "STEM Reform Outcomes through Communities of Transformation." *Change: The Magazine of Higher Learning* 48 (1): 30–8. https://doi.org/10.1080/00091383.2016.1121084.

Hill, L. B. 2020. "Understanding the Impact of a Multi-institutional STEM Reform Network through Key Boundary-Spanning Individuals." *The Journal of Higher Education* 91 (3): 455–82. https://doi.org/10.1080/00221546.2019.1650581.

Kezar, A. 2011. "What Is the Best Way to Achieve Broader Reach of Improved Practices in Higher Education?" *Innovative Higher Education* 36 (4): 235–47. https://doi.org/10.1007/s10755-011-9174-z.

Kezar, A. 2014. "Higher Education Change and Social Networks: A Review of Research." *The Journal of Higher Education* 85 (1): 91–125. https://doi.org/10.1080/00221546.2014.11777320.

López, N., D. L. Morgan, Q. R. Hutchings, and K. Davis. 2022. "Revisiting Critical STEM Interventions: A Literature Review of STEM Organizational Learning." *International Journal of STEM Education* 9 (1): 39. https://doi.org/10.1186/s40594-022-00357-9.

NAS (National Academy of Sciences), National Academy of Engineering, and Institute of Medicine. 2007. *Rising above the Gathering Storm: Energizing and Employing America for a Brighter Economic Future*. National Academies Press.

NASEM (National Academies of Sciences, Engineering, and Medicine). 2016. *Barriers and Opportunities for 2-Year and 4-Year STEM Degrees: Systemic Change to Support Students' Diverse Pathways*. National Academies Press. https://doi.org/10.17226/21739.

NCSES (National Center for Science and Engineering Statistics). 2023. *Diversity and STEM: Women, Minorities, and Persons with Disabilities 2023*. Special Report NSF 23-315. National Science Foundation. https://ncses.nsf.gov/pubs/nsf23315/report.

Reinholz, D. L., and N. Apkarian. 2018. "Four Frames for Systemic Change in STEM Departments." *International Journal of STEM Education* 5, article 3: 1–10. https://doi.org/10.1186/s40594-018-0103-x.

Reinholz, D. L., I. White, and T. Andrews. 2021. "Change Theory in STEM Higher Education: A Systematic Review." *International Journal of STEM Education* 8 (1): 37. https://doi.org/10.1186/s40594-021-00291-2.

Robinson, T. N. 2022. "The Myths and Misconceptions of Change for STEM Reform: From Fixing Students to Fixing Institutions." *New Directions for Higher Education* 2022 (197): 79–89. https://doi.org/10.1002/he.20429.

Segarra, V. A., C. Primus, G. A. Unguez, A. Edwards, C. Etson, S. C Flores, C. Fry, A. N Guillory, S. L Ingram, M. Lawson, R. McGee, S. Paxson, L. Phelan, K. Suggs, L. R. Vega, E. Vuong, J. C. Havran, A. Leon, M. D. Burton, J. L. Lujan, and M. Ramirez-Alvarado. 2020a. "Scientific Societies Fostering Inclusivity through Speaker Diversity in Annual Meeting Programming: A Call to Action." *Molecular Biology of the Cell* 31 (23): 2495–501. https://doi.org/10.1091/mbc.E20-06-0381.

Segarra, V. A., L. R. Vega, C. Primus, et al C. Etson, A. N. Guillory, A. Edwards, S. C. Flores, C. Fry, S. L. Ingram, M. Lawson, R. McGee, S. P. L. Phelan, K. S. E. Vuong, L. Hammonds-Odie, M. J. Leibowitz, M. Zavala, J. L. Lujan, and M. Ramirez-Alvarado. 2020b. "Scientific Societies Fostering Inclusive Scientific Environments through Travel Awards: Current Practices and Recommendations." *Cell Biology Education–Life Sciences Education* 19 (2): es3. https://doi.org/10.1187/cbe.19-11-0262.

12

ACCESS+: A Model for Supporting and Engaging STEM Professional Societies in Systemic Change

Heather E. Metcalf, Stephani Page, Learty L. Shaw, Lucas B. Hill, Bethany Farmer, and Ershela L. Sims

Over the past fifty years, research has repeatedly demonstrated social stratification across a variety of STEM opportunities, outcomes, and sectors—including educational access, achievement, degree attainment, mentoring, hiring, promotion, scholarly recognition, resource allocation, job security, sense of belonging, performance evaluation, leadership, and physical and psychological safety and well-being (Atherton et al. 2016; Barthelemy 2021, 2020; Blair-Loy and Cech 2022; Cech 2023, 2022; Charleston et al. 2014; Clancy et al. 2017; Griffin 2020; Harding 2006, 2015; Jaumot-Pascual et al. 2023; Kachchaf et al. 2015; McAlear et al. 2023; McGee 2020a, b; Metcalf et al. 2018; Montgomery and Page 2018; Nguyen et al. 2023; Ong 2023; Ong et al. 2011; Prescod-Weinstein 2021; Russell and Metcalf 2019; Torres 2023). Many of the approaches to addressing this have heavily relied on deficit models that aim to fix those who are affected (Denton et al. 2020; Margherio et al. 2023; McGee 2021, 2020; Ridgeway 2019). As Dr. Ebony McGee (2020) explains, students, faculty, and employees are left with a limited set of choices: "emulate or embody hegemonic values, navigate an environment that is hostile to their identities, or leave the field. Those who do attempt to navigate it are forced to cope with and manage their suffering, over and over again."

In this chapter, we present a rationale for working with STEM professional societies as the sites of change and share the ACCESS+ model that combines applied critical theoretical frameworks; tools for education, assessment, and reflection; and a community approach to empower change makers in STEM professional societies. We offer key lessons learned from the ACCESS+ approach with the goal of supporting others in making systemic and organizational changes in STEM professional societies and beyond.

Background and Context

STEM professional societies are uniquely positioned to foster systemic and culture reform in STEM because, as authorities for the disciplines they represent, they set the tone for acceptable research, policy, and practice; play a role in defining metrics of educational and career success and advancement; determine the values, norms, and standards for their fields; and act as gatekeepers to networks, publication and speaking platforms, career and educational opportunities, and more (Campbell-Montalvo et al. 2022; Segura-Totten et al. 2021; Hulede 2018; Huyck et al. 2021; Ingram et al. 2021; Leibnitz et al. 2022b; Segarra et al. 2020a, 2020b, 2020c). They also offer a cross-sector impact by serving academic institutions as well as academic-influencing organizations like accrediting agencies, federally funded research and development centers, corporations, and government agencies.

About ACCESS+

In 2017 the Alliance to Catalyze Change for Equity in STEM Success (ACCESS) was funded by the NSF's Division of Molecular and Cellular Biosciences to bring together minority-affairs committee leaders from five biological professional societies to collect, discuss, and share tools and resources to address the needs and concerns of racially minoritized members (Segarra et al. 2020c). Led by Dr. Verónica Segarra, the collaboration among the American Society of Biochemistry and Molecular Biology (ASBMB), the American Society for Cell Biology (ASCB), the American Society for Pharmacology and Experimental Therapeutics (ASPET), the Biophysical Society (BPS), and The Endocrine Society (TES) ultimately resulted in a cross-organization community of practice, a database of resources, and a travel award program (Segarra et al. 2020a) as well as promising practices for greater undergraduate student engagement (Primus et al. 2022) and speaker diversity (Segarra et al. 2020b) at ACCESS society conferences.

Building on the work of ACCESS, ACCESS+ is a partnership between two organizations, the Women in Engineering ProActive Network (WEPAN) and the American Association for the Advancement of Science (AAAS), and three initiatives, ACCESS, the ADVANCE Resource and Coordination (ARC) Network, and the Eddie Bernice Johnson INCLUDES Aspire Alliance. ACCESS+ was funded by the NSF ADVANCE program (#2017593) and the Burroughs Wellcome Fund (BWF). WEPAN, the home organization for ACCESS+, is a nonprofit educational organization that equips advocates with tools to create sustainable, systems-level changes that allow all in engineering to thrive. The AAAS is a global nonprofit organization focused on advancing science, engineering, and innovation throughout the world for the benefit of all. The ARC Network is the resource coordination hub for the NSF-ADVANCE program. It supports dissemination, community engagement, and capacity build for efforts related to the STEM professoriate. The Eddie Bernice Johnson INCLUDES Aspire Alliance is an NSF-funded initiative that leverages three change teams to support organizations and individuals in their change endeavors with the goal of national systemic change for STEM faculty.

ACCESS+ supports change agents from thirty-five STEM professional societies. The five original ACCESS societies serve as the first cohort of ACCESS+. The second cohort has fourteen societies and the third and most recent cohort has sixteen societies. Each participating society has up to three people from each organization who represent the society in ACCESS+ activities. These change agents come from a variety of positions in their societies, including paid staff, volunteers, committee members, and executive leadership. While the participating STEM societies joined ACCESS+ in three cohorts that were originally intended to last nine months to a year each, STEM societies from all three cohorts are still active in ACCESS+ activities.

ACCESS+ takes a community of practice approach to support STEM ecosystems (Costino 2018; Isler et al. 2021; Müller and Palmén 2022, pp. 1–24; Overstreet et al. 2020). We offer participants tools and resources to support their education, reflection, and growth; self and organizational assessment, action planning, and change implementation; and community development. In the following sections, we describe each of the elements of the ACCESS+ model in more detail.

Application

ACCESS+ participants join ACCESS+ through an application process. This application is designed to help the ACCESS+ team understand the context of the STEM society and its representatives, including their motivation for joining ACCESS+, the expectations they bring, the goals they have for their participation in ACCESS+, how they define and measure success, challenges they have encountered, and specific support they hope to receive. The ACCESS+ team uses this information to curate and situate resources within the participating STEM society's context. Throughout the life of ACCESS+ there have been two open application periods, and the team has been able to support all STEM societies who applied. Once a STEM society is accepted into the ACCESS+ program, their representatives spend about two months completing a self-assessment.

Self-assessment Tool

The ACCESS+ self-assessment tool (EEST) helps STEM professional societies assess and benchmark organizational change efforts, discern strengths, and identify areas for attention. The EEST is also intended to stimulate internal conversations and provide ACCESS+ participants with evidence-based promising practices and examples of actionable changes that can be made. In addition, the EEST allows participants across societies to draw on peer strengths to address challenges. The first version of the EEST (EEST 1.0) used a six-point Likert scale to evaluate a series of statements. STEM society participants:

1. Assessed their current practices within twelve STEM society functions, receiving a score that helped identify areas of improvement.
2. Inventoried if they regularly gather data within those functions.
3. Mapped out challenges and opportunities based on their score in item 1.

EEST 1.0 was a low-fidelity, spreadsheet version of the tool. ACCESS+ societies that completed the tool then received a report of the results along with recommendations, examples, and practices to support areas of needed improvement.

While helpful for inventorying the current state of organizational efforts within a STEM society, the EEST 1.0 did not provide guidance for internal data collection to assess areas for growth nor did it provide a great deal of open-ended questions designed to allow for deeper exploration and understanding of society change efforts. At the same time EEST 1.0 was launched, AAAS was developing a tool for STEM professional societies to use in self-assessment and action planning, modeled after the process used in the AAAS STEMM Equity Achievement (SEA) Change initiative for institutions of higher education (AAAS 2024). As the ACCESS+ team was revising the EEST 1.0, AAAS joined the partnership project, which allowed both organizations to learn from each another's approach and resulted in the EEST version 2.0. A low-fidelity version of the EEST is available on the ACCESS+ website via a Creative Commons license (CC BY-NC-SA).

The EEST 2.0 consists of four key components: (1) an organizational inventory and functional narrative; (2) an exploration of population characteristics; (3) progress, challenges, and priorities reflection; and (4) guidelines for internal data collection, which are adapted from the AAAS STEMM Professional Societies' Self-Assessment for Diversity, Equity, and Inclusion tool.

Part 1: Organizational Inventory and Functional Narrative

The organizational inventory and functional narrative of EEST 2.0 center on the twelve functional areas essential to a STEM society's overall operations, as identified in EEST 1.0. The twelve functional areas include: (1) governance and leadership; (2) membership; (3) meetings, conferences, and events; (4) professional development; (5) chapters and affiliates; (6) awards and recognition; (7) marketing and communication; (8) community outreach and engagement; (9) employment; (10) public policy and advocacy; (11) publishing; and (12) partners, sponsors, and vendors. EEST 2.0 streamlines the organizational inventory for each functional area to allow societies to assess their current practices more effectively while placing more emphasis on practices described in the functional narrative.

Organizational Inventory

The organizational inventory provides STEM societies with a six-point Likert scale to use in response to a series of statements that reflect concepts and practices within each of the twelve functional areas. The society representatives that complete the EEST use the following scale to assess the statements in each functional area:

- Not applicable—the society does not currently implement activities related to the functional area.
- Never—activities related to the functional area are minimal to nonexistent.

- Rarely—activities related to the functional area are in the ideation or planning stage.
- Sometimes—activities related to the functional area are being piloted or are in early implementation.
- Often—activities related to the functional area are regularly reviewed and improved.
- Always—activities related to the functional area have been established and are currently being implemented.

Functional Narrative

The functional narrative guides societies through a deeper exploration of each functional area to make a quality appraisal of the organization's current practices and leverage guidance provided in the AAAS tool. The functional narrative includes writing prompts for gathering qualitative descriptions of the work being done. Once the narratives are written, society representatives then appraise each of the functional areas using the following scale:

- Inactive—no actions are currently being taken to address the functional area being assessed.
- Passive—within the functional area being assessed, awareness of the need for organizational change exists and needs are known; however, very little action has been taken to implement change or address the needs of all populations.
- Transactional—within the functional area being assessed, awareness of the need for organizational change exists and needs are known; however, implementation is minimal, and the strategies employed are transactional, prescriptive, indirect, and minimally address the needs of all populations.
- Progressive—within the functional area, the need for organizational change is recognized. Actions are currently being taken to address these needs and processes, goals, and strategies are being established to sustain work in the area. However, there are still opportunities to strengthen these efforts to directly address the needs of all populations.
- Transformational—within the functional area, the need for organizational change is not only recognized but actively transformed through established policies and practices. These are (or will be) reviewed regularly to ensure continuous improvement and create more transformational approaches that directly address the needs of all populations.

Part 2: Population Characteristics

The original EEST tool provided guiding questions that allowed professional STEM organizations to inventory what data were regularly being collected that could be used to assess their change efforts. EEST 2.0 expands on this, providing societies with guided instructions for data collection and a model for exploring population characteristics across various compositional measures. The tool asks that organizations share (a) whether such data exists, (b) the frequency with which these data are collected, and (c) counts (n) of each dimension collected.

Societies are provided with a protocol and example table as a guide for collecting, disaggregating, and sharing data based on variables as well as creating tabular data sets for benchmarking and analyzing intersectionality across multiple dimensions. Organizations are encouraged to begin this descriptive analysis using available data and consider data usage and any additional data required to support organizational change efforts by using ethical and responsible data collection methods. Organizations using EEST 2.0 can leverage part 2 to identify gaps in available intersectional data that allows for internal discussions of which gaps need to be addressed and how, while maintaining the protection of personally identifiable information.

Part 3: Progress, Challenges, and Priorities

Part 3 of EEST 2.0 is intended to help professional STEM organizations (a) identify areas of success, (b) identify challenges, and (c) reflect on current organizational change strategies and priorities for the future. Organizations are also prompted to reflect on their process for selecting and building relationships with partners and vendors. Questions in part 3 are designed to guide discussions to identify changes that need to be made and the prioritization of those changes. Questions include:

1. What areas have been the organization's focus or priority and why?
2. If applicable, describe the progress made in the areas that the organization has focused on or prioritized, and the metrics used to measure this progress.
3. What accomplishments should the organization be proud of, if applicable?
4. What are the organization's main challenges in making progress in organizational change?

5. Describe (a) the approach used in the organization's change efforts, and (b) the insights gained from using that approach.
6. What opportunities have been identified based on the unique experiences of individuals?
7. What are the organization's priorities for the next 12–24 months?

Part 4: Guidelines for Internal Data Collection

EEST 2.0 provides an example guide for the internal collection and assessment of quantitative data to explore the intersectional demographic composition of members, volunteers, and staff. These data are not meant to be distributed outside of the organization as it is highly likely that representation in many categories will be so low that it will be impossible to protect personally identifiable information. Organizations are provided with a table to use as a guide for collecting and organizing demographic data about the population, composition of the members, volunteers, and staff of the professional organization or society over the past five years.

Ultimately, EEST 2.0 points organizations to the AAAS STEMM Professional Societies' Self-Assessment for Diversity, Equity, & Inclusion tool for more in-depth quantitative data guidance. This mitigates redundancy and reduces competition across initiatives with the same goal of transforming systems within our professional organizations to be more equitable and inclusive.

Upon completion of the EEST, ACCESS+ participants are provided with reports that synthesize the results, highlight areas of strength and opportunities for improvement, and offer additional resources, including a community of practice (CoP), an ACCESS+ convening, mini-grant opportunities, and a resource library to support them in their change work moving forward.

Community of Practice

Equipped with information from the EEST, ACCESS+ participants then engage in the ACCESS+ CoP throughout the year. Designed to provide peer, community, and expert support throughout the change process, the CoP serves as a space for participants to deepen their understanding and application of organizational change work, develop and gain feedback on their action plans, collaboratively solve implementation challenges, and share and exchange promising practices. The CoP supports ACCESS+ participants in becoming more aware of valuable resources, becoming more confident in navigating organizational change work, and developing their own expertise.

For the first cohort, we hosted CoP sessions approximately monthly, with an emphasis on developing action plans in response to participants' EEST reports and more deeply exploring the areas in the EEST that were most pressing to participating societies (e.g., meetings, conferences, events, and publication). We supplemented these sessions with workshops featuring experts on specific topics, like inclusive data collection.

As the second and third cohorts joined the CoP and we updated the EEST, we gained feedback from participants through surveys, polls, requests for feedback during the CoP, and direct communications. In response, we mapped out a standing schedule with topics and preparatory information. We also better aligned the CoP topics and invited experts with the revised EEST, created more space for in-depth action planning, incorporated discussions and learning opportunities about individual and organizational change and contexts, and placed a greater focus on how to adapt resources to those contexts. We worked with organizational change experts to develop and deliver a curriculum through the CoP that included more consideration of the various contexts around each individual's participation in ACCESS+ such as their own identities; roles, responsibilities, agency, and authority within their respective organizations; and the leadership and governance structure of their societies.

In our fourth year, we further leveraged the expertise and resources of the ACCESS+ team to expand our professional development curriculum and resources. This included engaging related initiatives our team members were involved in to bring in additional content, resources, and professional development opportunities to the CoP. This approach provided more in-depth learning to set the groundwork for cohort members to not only implement their action plans more effectively and meaningfully but also provide them with knowledge and skills to translate resources into practice with consideration of their specific organization's structure as well as their specific role within the organization. The CoP also prioritized cross-cohort engagement so the community could benefit from each other's experiences with organizational change.

Convening

The convening, one of the major goals of the ACCESS+ project, was included in the project to provide professional development and resources focused on ADVANCE policies and best practices for change-leaders. During initial convenings cohort members explored areas of strength and areas for development and explored how to adopt evidence-based policies and practices in their societies. As we received feedback the convening expanded to support STEM professional society, action–plan development focused on applying knowledge and resources to change activities within each society's context. Participants were also offered additional opportunities to connect and collaborate within and across cohorts and societies.

The first ACCESS+ convening took place in September 2021. During this two-day virtual convening, STEM professional society change leaders from the first cohort of ACCESS+ STEM professional societies reviewed their EEST results, engaged in force-field analysis (FFA) (Lewin 1951), and began developing action plans to address and strengthen the organizational change efforts within their societies. Participants also learned about the inclusive professional framework for societies (IPF: Societies) (Leibnitz et al. 2022a) and its application to their work. This convening also served as the kickoff for the ACCESS+ community of practice (CoP).

In the spring of 2022, we held our first in-person convening in Washington, DC. During this convening, cohort 1 participants reviewed the progress on their change efforts and identified next steps. Cohort 2 participants reviewed their EEST results and set priorities for their organizational change work. Cohort 1 revisited their original FFA and action plans from the first convening and adjusted or plotted a new area of focus. Cohort 2 developed their FFA and began drafting an initial action plan. Cohort members (1 and 2) also participated in an IPF: Societies workshop.

In the fall of 2022, we hosted another virtual convening to kick off cohort 3. During this virtual kickoff meeting—which took place on two consecutive days—STEM professional society change leaders from the third cohort reviewed their EEST results, engaged in FFA, worked on initial action planning (including evaluation strategies), and learned about the IPF: Societies. Participants used their EEST results to draft action plans to address areas of interest or concern. Societies were encouraged to share their action plans with other ACCESS+ cohort members to promote interorganizational knowledge exchange, especially around shared experiences related to solution identification or how to navigate the academic landscape.

Based on feedback from participants in the convening surveys, we determined that they wanted more time for action planning, networking, and feedback from the ACCESS+ team and other experts on the challenges they are facing. They also wanted to have more breakout sessions to discuss their EEST results within their society and with other societies. The evaluation survey respondents indicated that hearing new ideas from or having new conversations with others was the most helpful aspect of the convening. Some participants also indicated that more lead time for scheduling and better communication of details regarding the in-person convening would be helpful and allow them to make the meeting more impactful. In response, we updated the convening goals, expanded our schedule, and added workshop content. Workshops were designed to address four of the EEST frames: membership, programs, marketing and communications, and professional development.

The updated goals were to:

- Build participant awareness of areas of strength and areas for improvement within their society through discussions about EEST results.
- Help participants identify and understand the drivers and barriers of change within their society through discussions about EEST results and FFA.
- Lead participants in defining the actionable steps needed to achieve their goals through action-plan development.
- Allow participants to cultivate a peer support network and sense of community.
- Provide resources to support their goals and outlets for sharing their work.

The ACCESS+ 2023 convening was held in person in Durham, North Carolina. To allow ample time to address feedback from previous convenings, we extended the convening to cover a day and a half with a networking reception the evening before. During the convening, participants were able to advance action plans through two primary mechanisms. First, team members engaged participants in a sequence of activities to examine society and individual goals and missions, contextual factors, and supports and barriers to articulate change goals, key actions, implementation plan, and the beginning of an evaluation plan. In addition, we hired an ACCESS+ consultant who developed and delivered two workshops. In the second workshop, cohort members were provided with a framework for participants to address their goals within the functional areas of membership, professional development, marketing, communication, and programs. We repeated each of the workshops twice during the six weeks immediately following the convening to support the cohort members who were unable to attend the convening and to allow cohort members who wanted additional time to engage in professional development and action planning opportunities.

Based on evaluation data and feedback from cohort members, we specifically designed the 2024 convening to include opportunities for participants to gain some practical experience with several of the key functional areas. We developed a set of case studies to provide participants with an opportunity to apply the strategies learned in the CoP and other ACCESS+ professional development. Case studies were presented in two breakout sessions. In the first "Case Study Collab" session, team members presented case studies on governance and marketing and communications. During the second "Case Study Collab" session, participants worked through case studies on awards and recognition and annual meetings. Additionally, ACCESS+ members expressed interest in hearing about specific experiences that members of our project team had with several of the functional areas, so we kicked off the convening with a panel where members of the ACCESS+ team and one of the cohort three society members shared how they have navigated certain topics such as leadership buy-in as well as best practices and lessons learned. Last, participants were able to gain additional experience applying knowledge and resources to advance their action plans through the development of an organizational change toolbox. Cohort members engaged in a four-part process that included different ways to view organizations, described three different levels of change (individual, organizational, and systems), outlined multiple types of change levers that can be applied to organizational change efforts, and a review of those levers. In preparation for the session, one of the team members combined the case studies and organizational toolbox resources with all of the session activities and discussion questions in the form of a convening handbook, which was made available to all ACCESS+ members and included instructions on how to access the resources in the ARC Network Resource Library. To support the cohort members who were unable to attend the convening and to allow cohort members who wanted additional time to engage in the professional development provided during the convening, the case studies were presented during subsequent CoP meetings.

Mini Grants

Societies often express that there is a dearth of financial support to complete their work. To this end, ACCESS+ includes catalytic funding to support societies. Each year, cohort members can apply for mini grants to support projects related to their action plans. Societies that receive funding have nine months to complete their project and then they present their results at the next in-person convening. The projects funded by these mini grants cover a range of areas including leadership training, community building, deep-dive action planning, and inclusive communications.

During the first round, we awarded six mini grants at $2,000 each. During the second round, which was open to all three cohorts, we received a large number of applications. We decided to make the grants slightly smaller so that we could fund more projects and awarded fifteen mini grants.

During the most recent mini grant application cycle, we wanted to encourage societies to partner in their organizational change efforts. Therefore, we updated the criteria to require novel partnerships between two to three societies to create a deliverable associated with at least one of the EEST frames. We awarded two collaborative mini grants. One to develop best practices for an inclusive conference and the other for an emerging issues workshop on how STEM professional societies can support their higher education partners with their change work.

Resource Library

One way ACCESS+ facilitates and empowers positive change within STEM professional societies is by providing ACCESS+ participants with a robust set of resources that can be accessed on demand. The ARC Network Resource Library houses over 9,000 resources that range from peer-reviewed articles to guidelines, concept papers, policies, toolkits, conference presentations, training videos, and more. With a special collection specifically for STEM professional society contexts, the resources have been carefully curated, tagged, and organized by our community librarian in an ongoing effort to provide relevant and useful information to organizational change makers. In our curation efforts, we strive to include voices, perspectives, movements, disciplines, and methodologies often excluded and undervalued in STEM. ACCESS+ participants can also contribute resources that they have created or found helpful to the library, thus supporting their colleagues' change activities. Assistance from the librarian with searching and gathering resources is also available to all members by request.

Additionally, to remove as many barriers to information as possible, we have also begun creating LibGuides related to the ACCESS+ frameworks. Since searching for information is frequently nonlinear and iterative, we developed these guides to facilitate different approaches to gathering information; by providing a more welcoming look and feel, removing the need for account setup and passwords, and providing different types of consumable material such as videos, links, and social media accounts we are better able to accommodate the process of information gathering and the mental flexibility required to pursue a range of information resources (Sterner and Glasby 2023; Ugstad et al. 2023). Furthermore, to facilitate the discursive practice of scholarship created by scholars and academics engaging in continued research and conversation, we intend to further build out the LibGuides into a space for community engagement (Efird-Green et al. 2024). In this capacity members will be encouraged to share stories from their own experiences with the EEST, CoP, and change making at their institutions, including promising practices and lessons learned.

Findings

The team learned many valuable lessons through implementation of the ACCESS+ initiative over the past several years. As previously discussed, feedback collected by our evaluation team was used to make changes to our CoP, feedback, and other strategies in real time. Here we share lessons learned in several key areas to support others in making organizational changes in STEM professional societies and beyond.

Data Collection and Analysis

Throughout ACCESS+ we sought to illustrate a variety of data collection and analysis practices, including resources around these practices in the EEST, CoP, convening, and resource library. We found that guiding participants not just on how to collect and analyze data to inform their understanding, action planning, and decision-making, but also explaining why these practices are important has been critical to their implementation.

For example, we hosted a CoP workshop on community- and person-centered assessment that featured qualitative, quantitative, and mixed-methods researchers and evaluators. During the session, participants expressed a desire for actionable tips on how to create surveys of their members that help their organizations make decisions and set priorities. However, it was the why conversations throughout this and subsequent CoP sessions as well as individualized consulting with the ACCESS+ team that ultimately allowed the practices to crystallize for the participants. These ongoing discussions unpacked the underlying rationale for the recommended data collection and analysis practices and how they help societies make informed decisions and best serve all their members. For instance, during the workshop, we discussed that every item on a survey or interview script is a choice that conveys the society's values and priorities. Understanding the why empowered ACCESS+ participants to hold themselves accountable for the decisions they made and the priorities they set.

Leveraging Feedback

Gathering multiple forms of feedback from ACCESS+ participants allowed us to co-create tools, resources, and community as well as role model responsive and inclusive data collection practices. We obtained participant feedback through direct engagement with participants, the application form, periodic reports on focus groups and surveys from our external evaluation team, live feedback in CoP meetings, polls, pilot tests of the EEST, and more. This feedback has been incredibly valuable, equipping the ACCESS+ team with the information needed to make and communicate about adjustments along the way and better match resources to the contexts of our participants. For example, feedback about the CoP allowed us to better understand that our participants need test beds for putting ideas into practice, structured cross-cohort connection, self-efficacy and leadership development related to organizational change, and a better understanding of how to recognize and measure success. Being transparent about adjustments and why they are being made has contributed to a sense of trust, belonging, and investment in ACCESS+ from participants.

Peaks and Valleys

Our evaluation data have also taught us that leadership development is not a linear process. Rather, it is cyclical and full of peaks and valleys. ACCESS+ participants overwhelmingly rated themselves higher on self-efficacy metrics at the start of their ACCESS+ journeys than they did a year in, with one cohort reporting a 15 to 22 percent decrease in their belief that they can improve their professional societies by contributing to making change, developing a feasible action plan, implementing an action plan, and leading change. While this may seem like ACCESS+ is having the opposite of our desired impact, these participants simultaneously demonstrated a 46-percent increase in their understanding of how to improve their STEM professional societies. In other words, as participants engaged in ACCESS+, they deepened their knowledge about the scale and complexity of efforts to improve organizational efficacy and reassess their ability to make an impact. This was a pivotal moment where support was needed and for participants to be in a community so that they did not feel alone in facing the challenges that came with their organizational change efforts.

Action Planning

There were two main challenges that the team faced with action planning. The first challenge was to include the key constituents necessary within societies to make plans as efficacious as possible. Society representatives in ACCESS+ include those with different roles and positions within their organizations. With few exceptions, the society representatives that engage in ACCESS+ often did not occupy executive leadership positions and thus, like most change agents in higher education, work on the periphery or serve as volunteers within their organizations. The result is that many individuals in ACCESS+ did not hold decision-making authority to advance organizational change goals and strategies. The implication is that there are two levels of action planning, organizational and individual. Organizational action planning consists of making change at the organizational level. Individual action planning consists of people trying to advance change within an organization through their positions and relational influence. Thus, ACCESS+ programming evolved to engage participants on both levels to provide capacity building and resources related to action planning. This allowed processes and strategies to be repeated and replicated individually and have larger teams of society staff and leaders outside of convening and CoP spaces.

The second challenge was providing sufficient time to engage in action planning through ACCESS+ as indicated by the desires of members to expand that aspect over time. Even with additional sessions, the available professional development could never rise to the level of sustained organizational change consulting within the available grant funding. Thus, it was essential to revisit components of action planning over the course of several convenings and CoP sessions, both to provide needed support and to facilitate knowledge sharing between societies at all stages of planning and implementation. The goal was to provide accessible tools and resources that could be used to diagnose, strategically plan, implement, and examine results for ongoing growth and development.

Discussion and Reflection

It is important that STEM professional societies leverage their unique positions of authority to create STEM fields and spaces where everyone thrives. For the STEM professional society constituents reading this chapter, below are a few recommendations to consider when engaging in organizational change work.

Engage in ongoing professional development opportunities, join a CoP, or start your own. This work can often feel isolating and overwhelming. Being part of a community can not only help you feel less alone but also offer a rich space for the exchange of ideas, lessons learned, connection, collective impact, and more. Grounding the CoP in authentic and mutually beneficial relationships and practices allows community members to put into practice what they have learned, co-create together, and witness a living example of what is possible.

Inform your efforts with a comprehensive set of perspectives. Gather data and input from constituents across the professional society, taking care to hear from all members. As you are gathering data, intentionally work to build and maintain trust by being transparent about how you will use the feedback and follow up when you act on the feedback provided. Finally, create an action plan that includes a timeline and information about the resources needed to successfully implement the plan. Consider what resources, connections, and momentum are already available to you that you can leverage for change. Be sure to consider the challenges you might encounter and how you will address them.

Conclusion and Future Work

Over the past four years, ACCESS+ has supported thirty-five STEM professional societies, largely from the engineering and life science fields. Our team learned many valuable lessons in developing and growing the ACCESS+ initiative. In the future, we hope to expand this work to support additional STEM professional societies and fields. We also have plans to adapt the EEST to apply new contexts and focus areas for organizational change.

References

AAAS (American Association for the Advancement of Science). 2024. "STEM Professional Societies Self-Assessment." AAAS. https://www.aaas.org/programs/stem-professional-societies-self-assessment/resources.

Atherton, T., R. Barthelemy, D. Deconinick, M. Falk, and S. Garmon. 2016. *LGBT Climate in Physics: Building on an Inclusive Community*. American Physical Society.

Barthelemy, R. 2020. "LGBT+ Physicists Qualitative Experiences of Exclusionary Behavior and Harassment." *European Journal of Physics* 41 (6): 1–15. https://doi.org/10.1088/1361-6404/abb56a.

Barthelemy, R. 2021. *LGBT+ Faculty and Students in STEM: Heteronormativity, Exclusion, and Isolation*. ARC Network.

Blair-Loy, M., and E. Cech. 2022. *Misconceiving Merit: Paradoxes of Excellence and Devotion in Academic Science and Engineering*. University of Chicago Press.

Campbell-Montalvo, R., A. L. Putwen, L. Hill, H. E. Metcalf, E. L. Sims, J. W. Peters, A. N. Zimmerman, D. L. Gillian-Daniel, G. M. Leibnitz, and V. A. Segarra. 2022. "Scientific Societies Integrating Gender and Ethnoracial Diversity Efforts: A First Meeting Report from Amplifying the Alliance to Catalyze Change for Equity in STEM Success (ACCESS+)." *Journal of Microbiology and Biology Education* 23 (1): e00340-21. https://doi.org/10.1128/jmbe.00340-21.

Cech, E. 2022. "The Intersectional Privilege of White Able-Bodied Heterosexual Men in STEM." *Science Advances* 8 (24): abo1558.

Cech, E. 2023. "Engineering Ableism: The Exclusion and Devaluation of Engineering Students and Professionals with Physical Disabilities and Chronic and Mental Illness." *Journal of Engineering Education* 112 (2): 462–87.

Charleston, L., R. Adserias, N. Lang, and J. Jackson. 2014. "Intersectionality and STEM: The Role of Race and Gender in the Academic Pursuits of African American Women in STEM." *Journal of Progressive Policy and Practice* 2 (3): 273–93.

Chronicle staff. 2024. "DEI Legislation Tracker." *The Chronicle of Higher Education*. https://www.chronicle.com/article/here-are-the-states-where-lawmakers-are-seeking-to-ban-colleges-dei-efforts.

Clancy, K., K. Lee, E. Rodgers, and C. Richey. 2017. "Double Jeopardy in Astronomy and Planetary Science: Women of Color Face Greater Risks of Gendered and Racial Harassment." *Journal of Geophysical Research: Planets* 122 (7): 1610–23.

Cliburn, E. 2024. "Corporations Grapple with Growing Anti-DEI Pressure." *Insight into Diversity*. https://www.insightintodiversity.com/corporations-grapple-with-growing-anti-dei-pressure/.

Collins, P. H. 2015. "Intersectionality's Definitional Dilemmas." *Annual Review of Sociology* 41: 1–20.

Crenshaw, K. 1989. "Demarginalizing the Intersection of Race and Sex: A Black Feminist Critique of Antidiscrimination Doctrine, Feminist Theory and Antiracist Politics." *University of Chicago Legal Forum* 1989 (1): 139–67. https://chicagounbound.uchicago.edu/uclf/vol1989/iss1/8.

Crenshaw, K. 1991. "Mapping the Margins: Intersectionality, Identity Politics, and Violence against Women of Color." *Stanford Review* 43 (6): 1241–99.

Costino, K. A. 2018. "Equity-Minded Faculty Development: An Intersectional Identity-Conscious Community of Practice Model for Faculty Learning." *Metropolitan Universities* 29 (1): 117–36.

Denton, M., M. Borrego, and A. Boklage. 2020. "Community Cultural Wealth in Science, Technology, Engineering, and Mathematics Education: A Systematic Review." *The Research Journal for Engineering Education* 109 (3): 556–80.

Efird-Green, L., M. P. A. Social, E. Marion, D. Willis, J. M. Gierisch, and L. Corsino. 2024. "Designing and Implementing a Community-Engaged Research e-Library: A Case Study for Adapting Academic Library Information Infrastructure to Respond to Stakeholder Needs." *Duke University Libraries DukeSpace*. https://hdl.handle.net/10161/29269.

Griffin, K. 2020. "Looking beyond the Pipeline: Institutional Barriers, Strategies, and Benefits to Increasing the Representation of Women and Men of Color in the Professoriate." In vol. 35 of *Higher Education: Handbook of Theory and Research*, edited by L. Perna. Springer.

Griffin, K., and S. Museus, eds. 2011. *Using Mixed-Methods to Study Intersectionality in Higher Education*, no. 151 of *New Directions in Institutional Research*. Jossey-Bass.

Harding, S. 2006. *Science and Social Inequality: Feminist and Postcolonial Issues*. University of Illinois Press.

Harding, S. 2015. *Objectivity and Diversity: Another Logic of Scientific Research*. University of Chicago Press.

Hulede, I. V. 2018. "Preparing Students for Success in STEM: Role of Professional Societies." *Cell Biology Education–Life Science Education* 17 (3): es14.

Huyck, J. J., K. L. Anbuhl, B. N. Buran, H. J. Adler, S. R. Atcherson, O. Cakmak, R. T. Dwyer, M. Eddolls, F. El May, J.-T. Fraenzer, R. Funkhouser, M. Gagliardini, F. J. Gallun, R. L. Goldsworthy, S. Gouin, J. Heng, A. E. Hight, Z. Jawadi, D. Kovacic, R. Kumar, S. Kumar, S. R. Lim, C. Mo, L. S. Nolan, A. Parbery-Clark, D. V. Pisano, V. R. Rao, R. M. Raphael, L. A. J. Reiss, N. J. Spencer, S. J. Tang, V. D. Tejani, E. D. Tran, M. Valli M, G. D. Watkins, R. V. Wayne, L. R. Wheeler, S. L. White, V. Wong, M. C. Yuk, J. T. Ratnanather, and P. S. Steyger. 2021. "Supporting Equity and Inclusion of Deaf and Hard-of-Hearing Individuals in Professional Organizations." *Frontiers in Education* 6: 755457.

Ingram, J. C., A. E. Castagno, R. Camplain, and D. Blackhorse. 2021. "Role of Professional Societies on Increasing Indigenous Peoples' Participation and Leadership in STEMM." *Frontiers in Education* 6: 753488.

Isler, J., N. Berryman, A. Harriot, C. Vilfranc, L. Carrington, and D. Lee. 2021. "Defining the Flow —Using an Intersectional Scientific Methodology to Construct a VanguardSTEM Hyperspace." *Genealogy* 5 (1): 8.

Jaumot-Pascual, N., K. DeerInWater, M. Ong, and C. Silva. 2023. "'I Can Do Data for My People': Experiences of Giving Back for Native Undergraduates in Computing." *Cultural Studies of Science Education* 18: 879–909.

Kachchaf, R., L. Ko, A. Hodari, and M. Ong. 2015. "Career-Life Balance for Women of Color: Experiences in Science and Engineering Academia." *Journal of Diversity in Higher Education* 8 (3): 175–91.

Leibnitz, G. M., D. L. Gillian-Daniel, R. Greenler, R. Campbell-Montalvo, H. Metcalf, V. A. Segarra, J. W. Peters, S. Patton, A. Lucy-Putwen, and E. L. Sims. 2022a. "The Inclusive Professional Framework for Societies: Changing Mental Models to Promote Diverse, Equitable, and Inclusive STEM Systems Change." *Frontiers in Sociology* 6: 784399. https://doi.org/10.3389/fsoc.2021.784399.

Leibnitz, G. M., J. Peters, R. Campbell-Montalvo, H. Metcalf, A. Lucy Putwen, D. L. Gillian-Daniel, E. L. Sims, and V. A. Segarra. 2022b. "Refining a DEI Assessment Tool for Use in Optimizing Professional STEM Societies for Gender Equity." *Frontiers in Sociology* 7: 755372. https://doi.org/10.3389/fsoc.2022.755372.

Lewin, K. 1951. *Field Theory in Social Science: Selected Theoretical Papers* (D. Cartwright, Ed.). Harper.

Margherio, C., M. C. Horner-Devine, S. Mizumori, and J. Yen. 2023. "Connecting Counterspaces and Community Cultural Wealth in a Professional Development Program." *Race Ethnicity and Education* 26 (6): 772–92.

McAlear, F., T. Smith, K. DeerInWater, S. Koshy, R. Goins, A. Scott, & L. Novohatski. 2023. *State of Diversity: The Native Tech Ecosystem*. AISES and the Kapor Center.

McGee, E. O. 2020a. *Black, Brown, Bruised: How Racialized STEM Education Stifles Innovation*. Harvard Education Press.

McGee, E. 2020b. "Interrogating Structural Racism in STEM Higher Education." *Educational Researcher* 49 (9): 633–44.

Metcalf, H., D. Russell, and C. Hill. 2018. "Broadening the Science of Broadening Participation in STEM through Critical Mixed Methodologies and Intersectional Frameworks." *American Behavioral Scientist* 62 (5): 580–99.

Montgomery, B. L., and S. C. Page. 2018. "Mentoring beyond Hierarchies: Multi-mentor Systems and Models." Paper commissioned by the National Academies of Sciences, Engineering, and Medicine Committee on the Science of Effective Mentoring in STEMM. https://nap.nationalacademies.org/resource/25568/Montgomery%20and%20Page%20-%20Mentoring.pdf.

Müller, J., and R. Palmén. 2022. "Introduction: Reflecting on a Community of Practice Approach to Institutional Change for a Greater Gender Equality in R&I and HE–Policy and Practice." In *A Community of Practice Approach to Improving Gender Equality in Research*, edited by J. Müller and R. Palmén, 1–24. Routledge.

Nguyen, M., S. Chaudhry, M. Desai, K. Dzirasa, J. Cavazos, and D. Boatright. 2023. "Gender, Racial, and Ethnic Inequities in Receipt of Multiple National Institutes of Health Research Project Grants." *JAMA Network Open* 6 (2): e230855.

Olson, A., H. Hadero, and A. D'Innocenzio. 2024. "Assault on Diversity Efforts Has Chilling Effect on Companies." *Indian Country Today*, January 19. https://ictnews.org/news/assault-on-diversity-efforts-has-chilling-effect-on-companies.

Ong, M. 2023. *The Double Bind in Physics Education: Intersectionality, Equity, and Belonging for Women of Color*. Harvard Education Press.

Ong, M., C. Wright, E. Espinosa, and G. Orfield. 2011. "Inside the Double Bind: A Synthesis of Empirical Research on Undergraduate and Graduate Women of Color in Science, Technology, Engineering, and Mathematics." *Harvard Educational Review* 81 (2): 172–208.

Overstreet, N. M., L. Rosenthal, and K. A. Case. 2020. "Intersectionality as a Radical Framework for Transforming Our Disciplines, Social Issues, and the World." *Journal of Social Issues* 76 (4): 779–95.

Peters, J. W., R. A. Campbell-Montalvo, G. M. Leibnitz, H. Metcalf, A. Lucy-Putwen, D. L. Gillian-Daniel, E. L. Sims, and V. A. Segarra. 2021. "Refining an Assessment Tool to Optimize Gender Equity in Professional STEM Societies." WCER Working Paper No. 2021-7. University of Wisconsin–Madison, Wisconsin Center for Education Research.

Prescod-Weinstein, C. 2021. *The Disordered Cosmos: A Journey into Dark Matter, Spacetime, and Dreams Deferred*. Bold Type Books.

Primus, C., A. N. Zimmerman, A. K. Terovolas, K. F. Block, C. G. Brown, M. D. Burton, A. Edwards, C. M. Etson, S. C. Flores, C. Fry, A. N. Guillory, S. L. Ingram, R. McGee, D. L. Neely-Fisher, S. Paxson, L. Phelan, K. Suggs, L. R. Vega, E. Vuong, J. L. Lujan, M. Ramirez-Alvarado, and V. A. Segarra. 2022. "Scientific Societies Fostering Inclusivity in the Life Sciences through Engagement of Undergraduate Scientists." *Frontiers in Education* 7: 757816. https://doi.org/10.3389/feduc.2022.757816.

RAESC (Royal Academy of Engineering and the Science Council). 2020. *Diversity and Inclusion Progression Framework 2.0 for Professional Bodies*. https://raeng.org.uk/media/zjvnz0f4/progression-framework-guidance-2-0.pdf.

Ridgeway, M. 2019. "Deprogramming Deficit: A Narrative of a Developing Black Critical STEM Education Researcher." *Taboo: The Journal of Culture and Education* 18 (1): 157–67.

Russell, A., and H. Metcalf. 2019. *Transforming STEM Leadership Culture.* Association for Women in STEM. https://awis.org/wp-content/uploads/2019-Leadership-Report_FINAL_WEB.pdf.

Segarra, V. A., C. Primus, G. A. Unguez, A. Edwards, C. Etson, S. C Flores, C. Fry, A. N Guillory, S. L Ingram, M. Lawson, R. McGee, S. Paxson, L. Phelan, K. Suggs,

L. R. Vega, E. Vuong, J. C. Havran, A. Leon, M. D. Burton, J. L. Lujan, and M. Ramirez-Alvarado. 2020a. "Scientific Societies Fostering Inclusivity through Speaker Diversity in Annual Meeting Programming: A Call to Action." *Molecular Biology of the Cell* 31 (23): 2495–501. https://doi.org/10.1091/mbc.E20-06-0381.

Segarra, V. A., L. R. Vega, C. Primus, C. Etson, A. N. Guillory, A. Edwards, S. C. Flores, C. Fry, S. L. Ingram, M. Lawson, R. McGee, S. P. L. Phelan, K. S. E. Vuong, L. Hammonds-Odie, M. J. Leibowitz, M. Zavala, J. L. Lujan, and M. Ramirez-Alvarado. 2020b. "Scientific Societies Fostering Inclusive Scientific Environments through Travel Awards: Current Practices and Recommendations." *Cell Biology Education–Life Sciences Education* 19 (2): es3. https://doi.org/10.1187/cbe.19-11-0262.

Segarra, V. A., S. Blatch, M. Boyce, F. Carrero-Martinez, R. J. Aguilera, M. J. Leibowitz, M. Zavala, L. Hammonds-Odie, and A. Edwards. 2020c. "Scientific Societies Advancing STEM Workforce Diversity: Lessons and Outcomes from the Minorities Affairs Committee of the American Society for Cell Biology." *Journal of Microbiology and Biology Education* 21 (1): 1–7.

Segura-Totten, M., B. Dewsbury, S. M. Lo, E. G. Bailey, L. Beaster-Jones, R. J. Bills, S. E. Brownell, N. Caporale, R. Dunk, S. L. Eddy, M. E. García-Ojeda, S. M. Gardner, L. E. Green, L. Hartley, C. Harrison, M. Imad, A. M. Janosik, S. Jeong, T. Josek, P. Kadandale, J. Knight, M. E. Ko, S. Kukday, P. Lemons, M. Litster, B. Lom, P. Ludwig, K. K. McDonald, A. C. S. McIntosh, S. Menezes, E. M. Nadile, S. L. Newman, S. D. Ochoa, O. Olabisi, M. T. Owens, R. M. Price, J. W. Reid, N. Ruggeri, C. Sabatier, J. L. Sabel, B. K. Sato, B. L. Smith-Keiling, S. D. Tatapudy, S. J. Theobald, B. Tripp, M. Pradhan, M. J. Venkatesh, M. Wilton, A. M. Warfa, B. N. Wyatt, and S. A. Raut. 2021. "Chronicling the Journey of the Society for the Advancement in Biology Education Research (SABER) in Its Effort to Become Antiracist: From Acknowledgement to Action." *Frontiers in Education* 6. https://doi.org/10.3389/feduc.2021.780401.

Sterner, J., and H. Glasby. 2023. "LibGuides as Outreach: Creating Research Guides about and for Diverse Communities." *Pennsylvania Libraries: Research and Practice* 11 (2): 6–16. https://doi.org/10.5195/palrap.2023.285.

Torres, L. E. 2023. "Sobreviviendo sin Sacrificando (Surviving without Sacrificing)—An Intersectional DisCrit Testimonio from a Tired Mother-Scholar of Color." In *Enacting Disability Critical Race Theory: From the Personal to the Global*, edited by B. A. Ferri, D. J., Connor and S. A. Annamma, 623–36. Routledge.

Ugstad, J., J. Valenzuela, and T. Spence. 2023. "Technical Services, Social Justice, and LibGuides: A Model for Impact." *Technical Services Quarterly* 40 (2): 45–58.

Afterword: The Importance of the Distorting Lens

Lydia Villa-Komaroff

How humans treat other humans is a function of how our brains have evolved as well as the effect of historical events on our attitudes to individuals from other races and cultures. That is the insight developed by many over the last fifty years and perhaps summarized most clearly by Jennifer L. Eberhardt in her book "*Biased: Uncovering the Hidden Prejudice That Shapes What We See, Think, and Do*" where she writes, "Implicit bias is a kind of distorting lens that's a product of both the architecture of our brain and the disparities of our society" (Eberhardt 2020). Recognizing that reality is important when determining how an organization should approach increasing diversity in its ranks.

Historians, anthropologists and economists have documented that in virtually all societies throughout recorded history some populations have been treated inequitably by the population in power, often the majority population, and women have generally not been granted rights accorded to men. Furthermore, the existence of these inequities was visible to all, which leads to the normalization of these inequities within the culture.

Why do these inequities persist? In this essay, I argue that the sciences of psychology, sociology, economics, and neuroscience have taught us that our brains evolved to make very rapid decisions based on what our accumulated experience tells us is normal. This allows us to ignore much of the overwhelming and constant flow of information received by the brain from the senses. However, this also means that we very rapidly recognize humans clearly different in some respect from ourselves and make erroneous assumptions about their abilities or the threat they may pose to us.

The chapters in this book report case studies from various professional societies attempting to increase diversity in their ranks. The underlying assumption of the chapters—which I share—is that most professional societies and organizations in our society do not reflect the diversity of our society because they are not equally welcoming or accessible to all groups and that our society and its organizations will benefit from solving that issue.

Every professional society is a sum of the values and attitudes of the discipline and individuals it represents. Consequently, the approach of each society to increase diversity in its ranks is more likely to be successful when the leaders of a society who are developing specific actions and policies take into account the experiences and values of its members. This book describes some of the activities and actions taken or underway by various scientific societies to increase participation in the organization and the discipline represented.

Making a case for the value of diversity, equity, and inclusion (DEI) is difficult at present due to deep divisions in the United States and because much of the vocabulary traditionally used to advance diversity has been co-opted and inverted by vocal, determined, and organized opponents. Arguments of justice or fairness can be especially fraught as opponents have declared that such activities represent reverse discrimination that results in disadvantaging white and some Asian individuals.

The "Hard" Sciences and "Soft" Sciences

The *life sciences* (biology) and *physical sciences* (physics, chemistry, astronomy, earth science, and the mathematical sciences that support them) are sometimes called the "hard" sciences, whereas the social sciences (e.g., sociology, psychology, history, economics, political science, anthropology, linguistics) are sometimes called the "soft" sciences. This pejorative distinction flatters "hard" scientists like me, many of whom know little about the knowledge accumulated by the "soft" scientists and how that knowledge may be valuable for those in the "hard" sciences.

In particular, knowledge from the social sciences, in conjunction with knowledge from the neurosciences, helps explain the underrepresentation of women and certain racial and ethnic groups in STEMM (science, technology, engineering, mathematics, and medicine). Those underrepresented groups and races include people who identify as Black, African American, Hispanic, Latino,[1] Asian, South Asian, and Indigenous or Native American.[2]

Scientists seek data. Social scientists collect data and use rigorous data analytic techniques. Exposure to those data and techniques—a data-based introduction to the science and history that underlie the current underrepresentation of certain groups and races—can increase understanding of the value of and the reasons why DEI efforts are necessary. Such understanding may be of use when building consensus.

In this essay, data from various fields are summarized and references provided that may be useful when responding to questions about the value and the need for explicit policies and actions to ensure that all individuals are provided opportunities to reach their full potential.

I argue that the reason our society, including our professional societies, are not fully diverse, equitable, or inclusive is a function both of how our brains have evolved, and how past attitudes and behaviors have influenced what is socially acceptable. Our brains perceive skin color and gender and make unconscious associations before conscious thought can occur, which can result in errors in judgment.

The Human Brain Makes Systemic Errors in Judgment

The study of human decision-making arose in part as psychologists and economists sought to understand financial decision-making. Many economic studies assumed that humans would make rational decisions to optimize their economic well-being. In the 1950s Nobel laureate Herbert A. Simon concluded that rational choice was an unrealistic description of how people make decisions in the real world, in part because it was so often impossible for the individual to know what the best decision might be (Campitelli and Gobet 2010; Goldstein 2015). The idea was more fully developed in the early 1970s by Tversky and Kahneman (1974) who used the term "heuristics" to describe the mental shortcuts used to make intuitive decisions. They developed experiments that demonstrated that rapid decisions made by individuals were strongly influenced by how a question was phrased.

As a result of their studies, Tversky and Kahneman[3] described two ways of thinking: one results from "one conscious thought leading in an orderly way to another" and a second that occurs without our conscious awareness. "The mental work that produces impressions, intuitions, and many decisions goes on in silence in our mind" (Kahneman 2013). While this intuitive knowledge is often correct, it can result in errors of judgment.

The first mode of thinking, called "System 1" by Tversky and Kahneman, operates automatically and quickly. Little or no effort is required on the part of the individual, there is no voluntary control over this system, and it is active continuously. These are intuitive responses, sometimes described as a "gut feeling." System 1 includes innate survival skills and allows us to perceive the world around us more rapidly than we can consciously assess it, which minimizes the time and energy of cognitive work. Because we recognize familiar objects without having to think, we can perceive novel objects and people as possible threats. Some mental activities can become fast and automatic with prolonged practice or exposure. We do not need to think about our trip to work or home after we learn the route. Professional musicians do not think about how they should hold their instrument as they play, it has become automatic (Kahneman 2013).

System 1 uses associative thinking: causes are linked to their effect (virus → cold), things are linked to their properties (lime → green), and things are associated with their categories (cherry → fruit). Once an idea has been activated, a rich network of related ideas will be invoked, and we will be aware of only a few of them. For example, "dog" will evoke images of dogs and either warm memories or fear, depending on your experience.

The second mode, System 2, allocates attention to effortful mental activities. It is orderly and deliberative; it requires concentration and attention. System 2 requires substantially more energy than system 1. Brains use more glucose when System 2 is at work, for example preparing taxes or a grant application.

Human Brain Development Continues after Birth

Input from the senses is perceived by the brain beginning about seven weeks of gestation in utero as measured by an increase in fetal heartbeat rate when an input is new or unusual. Humans in utero can touch, hear, taste, and see, and they react to stimuli around them, including touch (by another sibling conceived at the same time), their mother's voice, and light. A baby is born with the ability to recognize sounds they heard in utero, including the voice of mother and other family members, as well as recurring sounds such as a passing train or music (Reid et al. 2017; Donovan et al. 2020; Padilla and Lagercrantz 2020).

Neuronal plasticity allows us to adapt to changes in our environment. The brain activity that determines behavior depends on both genetic and developmental processes and environmental factors, such as learning through experience (Kandel 1991). Much of human brain growth and development occurs after birth under the influence of both the internal and the external environments (Klawans 2000; Rosenberg 2021). The human brain reaches its final size in early adolescence but does not reach full maturation until the mid-twenties (Nolte 1988; Giedd 1999; Arain 2013). The brain is not a static structure but changes as we learn and age (Cozolino and Sprokay 2006). A great deal of learning, some we are aware of and some we are not, appears to allow the brain to determine what events or characteristics occur reliably together (Kupfermann 1991). Events and characteristics of people that appear to be associated in an individual's environment and culture during infancy and childhood are often dealt with by System 1. While these rapid, automatic responses are frequently useful, they can also result in a biased response (Kahneman 2013; Banaji and Greenwald 2013). Several imaging studies provide support for neural activity in specific regions of the brain that correlate with rapid, often unconscious prejudicial reactions to people from historically excluded groups (Amodio 2014; Eberhart 2005; Golby et al. 2001).

The Ubiquity of Implicit Bias

It must be emphasized that all humans have implicit or unconscious[4] biases; they result from the way the human brain has evolved. Therefore, when cultural actions and history emphasize certain associations, many members of that culture will share some of those biases, even those against groups with whom they most closely identify (Banaji and Greenwald 2013; Eberhardt 2020). In a complex, multicultural society such as ours, we can develop associations that lead System 1 to make mistakes.

Culturally shared mental content of which we are unaware results in associations that can guide our behavior without our awareness. Our attempts to align our behavior with what we believe about ourselves can be unsuccessful. Two examples follow:

> I explained that some years ago my son Everett and I were on a plane. He was five years old, wide-eyed, and trying to take it all in. He looked around and saw a Black passenger. "Hey, that guy looks like Daddy." I looked at the man and truth be told he did not look anything like Daddy ... I couldn't help but be struck by the irony: the race researcher having to explain to her own Black child that not all Black people look alike...I gathered my thoughts and turned to my son, prepared to lecture him ... But before I could begin, he looked up at me and said,

> "I hope that man doesn't rob the plane" … "Why did you say that," I asked … Evertt looked up at me with a really sad face and said very solemnly, "I don't know why I said that, I don't know why I was thinking that" (Jennifer Eberhart in *Bias: Uncovering the Hidden Prejudice That Shapes What We See, Think, and Do*, 2020).

> We put down briefly in Khartoum, where we changed to an Ethiopian Airways flight to Addis. Here I experienced a rather strange sensation. As I was boarding the plane, I saw that the pilot was Black. I had never seen a Black pilot before, and the instant I did, I had to quell my panic. How could a Black man fly an airplane? (*Long Walk to Freedom: The Autobiography of Nelson Mandela*, Mandela 1995)

It's a Big Data Problem

The human mind constantly receives more data through each of our senses than the brain can consciously process, so mechanisms to ignore large amounts of incoming sensory data are essential. These mechanisms most likely arose during the evolution of the brain and nervous system. The human brain processes about 11 million bits of information per second but is only able to consciously manage 40–50 bits of information per second (Wu et al. 2016; Ramuthi 2024). Because the human brain receives environmental influences during natal and postnatal development that continue as the brain develops, learned associations can drive unconscious reactions to events or people (Banaji and Greenwald 2013; Eberhardt 2020).

Because we receive data constantly, our conscious actions may reflect perceptions that occur before we are consciously aware of that information. The amygdala of the brain is activated by a 30-millisecond exposure to a Black or white face, while the conscious perception of race takes about ten times longer. Both race and gender are perceived very rapidly but response to race, especially to Black faces, is stronger (Contreras et al. 2013; Eberhardt 2005; Ito and Urland 2003). In other words, we cannot be color blind because we perceive skin color and gender before we can consciously think about it. That is the way we are wired.

Our brain receives input from the senses, but what we perceive depends on how the brain processes that information. In the case of vision, it receives information about light, shading, sharpness, and apparent size to interpret the sensory information. Optical illusions occur when our brain interprets the input from our eyes in a way that does not accurately reflect reality. Similarly, implicit (or unconscious) biases stem from associating things that are not necessarily related. Some associations are learned unconsciously as we are exposed to aspects of our society such as family attitudes, movies, TV, peers, news, advertising and who we see in various roles (teacher, scientist, engineer, judge, lawyer, police). Some of these unconscious associations may be counter to what we believe about ourselves, and we are often unaware that we have made them.

The following associations are frequently made by many members of American society: dark Hispanic woman → maid; dark Hispanic man → laborer; and Black man → criminal. Unconscious, or implicit, bias is the process of making an inference based on a stereotype or attitude toward categories of people without being aware of doing so. Such unconscious biases permeate our culture and can influence our behavior without our awareness. The basis of many DEI training programs is the premise that identifying and understanding errors of judgment may limit the detrimental effects of implicit biases.

The Myth of Merit as the Sole Criteria for Evaluation in STEM

There is ample evidence that we act on implicit assumptions before we assess merit. A standard experiment is to expose a set of people to applicants with identical credentials but who differ in race, ethnicity, or gender (Pager and Shepherd 2008). One study showed that individuals named Emily or Greg received 50 percent more callbacks for jobs than did individuals named Lakisha or Jamal, despite having identical credentials for a job (Bertrand and Mullainathan 2003).

In 2012, Moss-Racusin and her colleagues performed an experiment to test whether faculty in biology and physics might display implicit bias against women students (2012). They undertook this study to directly address several articles arguing that gender discrimination did not exist in these fields because women prefer fields outside of STEMM, assume most family care responsibilities, and because of the rigorous training scientists receive in the objective evaluation of data (Berezow 2011; Ceci and Williams 2011; Ceci et al. 2011; Dickey Zakaib 2011). The faculty participants were recruited from the biology, chemistry, and physics departments of six highly regarded departments from large research universities in the United States whose undergraduate and graduate students tended to enter careers in science. The authors asked tenure-track faculty to evaluate the curriculum vitae (CVs) of potential candidates for a job as laboratory manager, an entry-level position in a science career.

This was the first large study to directly address the existence of implicit bias against women among science faculty from prestigious academic institutions; it came to similar conclusions as many previous smaller studies of gender, racial, and ethnic bias in multiple professional settings.

The authors used well-established and rigorous methods in the study and provided sufficient information in the text and in supplemental materials for others to evaluate the methods used.[5]

The participating faculty were randomly assigned a set of materials (explanation of study, CV with demographics and background, and excerpts from student statements and recommendation letters) that were identical with respect to credentials. However, half of the faculty received a CV with a masculine name, and the other half received the same CV with a feminine name. The materials sent to faculty were adjusted for differences in the job of a lab manager in different fields. The credentials described a good but not outstanding student.[6]

The hypotheses tested were:

1. "Science faculty's perceptions and treatment of students would reveal a gender bias favoring male students in perceptions of competence and hireability, salary conferral, and willingness to mentor."
2. "Faculty gender would not influence this gender bias."
3. "Hiring discrimination against the female student would be mediated (i.e., explained) by faculty perceptions that a female student is less competent than an identical male student."
4. "Participants' preexisting subtle bias against women would moderate (i.e., impact) results, such that subtle bias against women would be negatively related to evaluations of the female student, but unrelated to evaluations of the male student."

Faculty participants were asked to indicate on a numerical scale (a) their evaluation of the competence of the individual; (b) how likely they would be to hire the individual; and (c) how willing they would be to mentor the individual. They were also asked what they would offer in compensation.

The male applicant was judged more competent, a more attractive hire, a person faculty would be more willing to work with and mentor, and a person who should receive a higher salary. Women faculty were just as likely as men faculty to prefer the male job applicant. Furthermore, using a standard instrument, the authors found that a preexisting subtle bias against women was associated with less positive evaluation of women but not men (Moss-Racussin et al. 2012).

One study (Li and Koedel 2017) compared the overall composition of tenure-track faculty during 2015–16 academic year in three STEM (biology, chemistry, and economics) and three non-STEM departments (educational leadership and policy, English, and sociology) at forty selective public institutions across the United States. Black and Hispanic faculty were underrepresented in the STEM departments relative to their representation in the US population, while Asian and white faculty were over-represented. Disparities were not as great in the non-STEM departments. Furthermore, Black and Hispanic faculty had lower salaries than their white and Asian counterparts of equivalent rank and responsibility.

When faculty representation was examined by rank, there was more diversity at the assistant professor level than at associate and full professor levels, except for Black faculty, who were as underrepresented in the assistant professor level as they were at higher ranks. They next compared the representation of assistant professors to the proportion of PhDs from historically excluded groups produced by the top fifty research intensive universities in the United States. Black assistant professors were underrepresented in biology and chemistry, close to parity in economics and education leadership and policy, and overrepresented in English and sociology. Hispanic assistant professors are underrepresented in biology, education leadership and policy, and English; reach parity in chemistry; and overrepresented in economics and sociology. Asian assistant professors are overrepresented in every field examined except sociology, where they are underrepresented. White assistant professors are underrepresented in sociology, more underrepresented in economics than Black assistant professors but at or near parity in the other fields (Li and Koedel 2017).

A more recent study (Masters-Waage et al. 2024) evaluated 1,571 faculty promotion and tenure decisions from five US universities. While there was no difference in the percentage of departmental negative votes or unanimous votes, Black or Hispanic faculty received 7 percent more negative votes and were 44 percent less likely to receive unanimous votes from college promotion and tenure committees than white or Asian faculty. Faculty from these historically excluded groups were also more harshly judged for below average h-indexes than white or Asian individuals with the same h-indexes. These results were more negative when the individuals were Black or Hispanic women, but the results were mitigated when letters from external reviewers contained a higher proportion of words related to scholarship as determined using a well-regarded linguistic word count software.

Changing Minds vs. Changing Behavior

There is ample evidence for the value of diversity. Teams made up of people with different backgrounds and diverse points of view generate more approaches to problems than less diverse teams if team members have mutual respect (Smith-Doerr et al. 2017). Companies with a more diverse executive team are more likely to be in the top quartile of financial performance (Dixon-Fyle et al. 2020). Until about 2016, leaders of most corporations as well as government, military, and academic institutions generally agreed that diversity was a laudable goal and yet progress remained slow.

The National Science Foundation first published data on historically excluded groups in 1977. In 1980, the US Congress required that such a report be published biennially. The requirement of regular reports of such data recognized the widespread consensus that increased diversity in STEM is a laudable goal for government, universities, and companies. Although substantial resources were devoted to developing practices that might encourage students from historically excluded groups to enter and stay in STEM, the rate of change has been too slow to fill all the needs of the nation.

In 1978 the US Supreme Court decision in *Regents of the University of California v. Allan Bakke* (438 US 265) established that colleges could use race as one factor to achieve "educational benefits"; the Court did not say that race could be considered in order to remedy past discrimination (Carlton 2023). After the June 2023 US Supreme Court decision to disallow the use of race in college admissions, many colleges and universities came under pressure to discontinue activities that are targeted to recruit and nurture these populations (Gretzinger 2024). Following the 2023 ruling, early data indicate that entering classes at some highly selective universities have had fewer minority students than in the years immediately prior (Knox 2024). For the colleges and universities that seek to increase the fraction of new students from historically excluded groups, the best mechanisms for doing so are not clear. There was also a de-emphasis of DEI activities in the business world.

In fact, implicit bias training and antiracism training have not yet been shown to be effective in changing minds or behaviors (Devine and Ash 2022; Lu 2024).

There is substantial evidence that implicit bias training can have a transitory effect, but the effect appears to wane over time. Antiracism training can result in a clearer understanding of the structural barriers in our society, but it can also lead to backlash, especially among white and Asian populations, if the training is perceived as accusatory. Coupled with higher prices, wages that have not kept up with inflation, stove-piped news sources, unchecked social media, and a widening wealth gap, that backlash can have far-reaching effects on education and efforts to identify and nourish all the talent available in our society.

To increase the rate of progress, many institutions initiated training, first in implicit bias, then adding or substituting antiracism training. Some people, including many scientists trained in the life sciences or physical sciences, are angered by mandatory training to reduce racism or structural barriers to increased diversity. They perceive an accusation that they are racist and biased and may consider such training an additional unnecessary regulatory burden to be completed as rapidly and mindlessly as possible. Even if they have heard the terms "diversity," "equity," and "inclusion," they may not have a deep understanding of what these terms mean, the history behind them, or knowledge of the lived experience of people from these groups.

Talent is equally distributed; access to opportunity is not. The optimal approach would be to address the problems at the K–12 level of education so that all students could leave high school with strong basic skills, a sense of their talents, and the paths open to them. In those circumstances, children would be exposed to a variety of possibilities and could try them out. Many children start playing some sports early, in these sports the most talented individuals rise to the top and professional and Olympic teams have a deep pool of talent to draw on as they seek the best for world competition.[7] Exposing all children in K–12 to activities in STEM to create a deep pool of talent has never been among the highest priorities of government and is unlikely to be for the foreseeable future. However, there are many organizations dedicated to early STEM exposure and they can have a profound influence on children. Examples include the Science Club for Girls, American Rocketry Challenge, and Project Lead the Way. Most science museums have programs for children. Studies of successful programs and how to effectively bring them to the attention of teachers and parents are needed.

The most effective approach in higher education, professional organizations or corporations appears to be changes in policies and procedures that help to prevent the prejudgment of individuals. Such an approach requires thoughtful leadership, strategic planning, implementation with specific goals and milestones, transparency, accountability, and inclusion of all affected populations. Such changes are slow, but when successful, can be transformative. At best, policies and procedures that benefit people from historically excluded groups benefit all members.

One outstanding model is Georgia State University (GSU) (Gumbel 2020). The story of GSU illustrates both what success in a university setting looks like and how difficult and lengthy the road can be. In this case GSU went from a school with high dropout rates,[8] gaps in Black and white student retention,[9] and low graduation rates,[10] to a school with no gaps between white and Black students in these measures. The majority of the GSU students fall into one or more of the following categories: Black, Asian, Hispanic or Latino, first generation, and Pell grant eligible. The school tracks student performance and provides advising and other interventions to students identified as at risk by predictive analytics that provide real-time information on student performance. Incoming students are provided with more information on careers and starting salaries related to their interests. The first-year curricular experience has been extensively modified to provide incoming students with more flexibility so students can change direction without starting all over again. As a way of eliminating non-academic barriers to success, financial aid is provided to students who are both in need and are likely to succeed given necessary aid.

A second outstanding model is Duke University, which has seen an increase in diversity of highly qualified members of historically excluded groups among students and, more recently, among faculty. In this case, the institution involved the entire university in activities meant to make students feel part of the University. These activities included team-building experiences before classes began and matching incoming students with both faculty and student mentors. The trios meet on a regular basis during the year for both science-related activities such as discussing seminars, and nonscience activities such as meals together (Jones et al. 2022). For faculty, the approach is a series of programs that support all faculty in critical areas of professional life. These workshops and programs start with new faculty orientation. As a career advances, coaching, seed grants for faculty-led initiatives to foster a sense of community within an academic unit, workshops in requirements and expectations for promotion for both tenure-track and non-tenure track faculty are available (https://facultyadvancement.duke.edu/faculty-development/).

A third example is SEA[11] Change, (STEMM Equity Achievement) a program housed at the American Association for the Advancement of Science (AAAS), SEA Change provides a self-assessment process based on evidence-based change management that institutions can utilize to evaluate their practices and institute changes in practices and policies to build an environment where fostering success for all and scientific excellence are inseparable. Members of SEA Change have access to resources, courses, and training that include a template for self-evaluation, as well as on-demand courses at reduced cost, support from SEA Change staff through site visits and virtual meetings. Members have access to a shared website where they can share and discuss best practices and strategic approaches. Finally, SEA Change recognizes institutions' commitment and progress with awards (https://seachange.aaas.org/).

While there are best practices that can be learned from the examples in this book and in the literature, no one set of approaches will work in all settings. Every institution has unique features, so the best approach for a particular institution must come from within the institution and be molded to fit the institution as times and circumstances change.

While there is good reason to be pessimistic about the political climate for the next decade, it is important for organizations to identify and nurture talent available in all populations within the limits of the law. Those of us in states that encourage such activities must support our colleagues where this work has come under fire. Academic studies must continue, perhaps with a greater focus on how all individuals—in an organization or in society at large—benefit from efforts to achieve DEI.

The "distorting lens" cannot be eliminated. Instead, a corrective lens must be put in place. The neurological processes that underlie implicit bias will not change and societal attitudes change at a glacial pace. Humans will always receive more sensory data than can be consciously processed. A recent evidence-based analysis of research suggests that individual actions at the local level can give rise to collective action that lead to procedures, processes, and practices that ameliorate our implicit tendencies (Dasgupta 2025).

Acknowledgments

I thank Florence Bonner for pointing out to this molecular biologist the importance of the contributions that social scientists have made to understanding human behavior. I am grateful to colleagues from the National Science Foundation's (NSF) Committee on Equal Opportunity in Science and Engineering; the NSF Advisory Committee for Social, Behavioral and Economic Science; the National Academies of Science, Engineering and Medicine standing Committee on Women in Science, Engineering, and Medicine; and all the other colleagues who have commented on my work as it developed and who have studied and worked toward DEI in the sciences and business over the years. This work was accelerated by the award of the 2016 Elting E. Morrison Prize from the MIT Program in Science and Technology. I thank Anthony L. Komaroff and Ada Hagan for their insightful and thoughtful editorial suggestions.

Notes

1 I use Latino instead of Latine or Latinx as it is the term favored by most individuals in this group.

2 It must be kept in mind that each of these populations is made up of subpopulations whose participation may differ.

3 Kahneman received the Nobel Prize in Economics in 2002; Tversky did not share the award because he died in 1996.

4 Psychologists and other social scientists have adopted the term "implicit bias" because they found conceptual difficulties when the term "unconscious" was used (Greenwald and Banaji 2017).

5 The departments were matched for size and prestige and represented three different geographic areas. One private and one public institution were selected from each region. The demographics (field, gender, and race/ethnicity) of the 127 faculty who completed the study reflected the average demographics of the sampled departments and similar departments nationally. Distribution by rank was also noted. Demographic variables were not associated with any responses. A power analysis indicated the sample size was greater than needed to assure the detection of moderate effects as well as sufficient to test four hypotheses directed at determining whether natural science faculty from prestigious institutions harbored implicit gender bias.

6 The subjects and recruitment strategy, the application materials, dependent variable scales, analytic strategy, mediation analysis, moderation analysis, and the text of the cover story provided to participating faculty, and the application CV. http://www.pnas.org/lookup/suppl/doi:10.1073/pnas.1211286109//DCSupplemental/pnas.201211286SI.pdf.
7 This idea was first proposed by Elma González, PhD, a plant cell biologist, Professor Emeritus at the University of California, Los Angeles (UCLA), and the first Mexican American woman to receive a PhD in a natural science (from Cornell University). She was a faculty member at UCLA and (along with Clifton Poodry, PhD) was instrumental in the expansion of the Maximizing Access to Research Careers program and others like it to majority serving institutions.
8 Percentage of students who leave before completing a degree.
9 Percentage who stay enrolled in a degree program; first-year retention is a common measure used.
10 Percentage of first-time, full-time undergrads who complete their program in a given time.
11 The name was inspired by the poem The Cure of Troy by Seamus Heaney. "Once in a lifetime / the longed for tidal wave / of justice can rise up, / and hope and history rhyme. // So hope for a great sea-change / on the far side of revenge. / Believe that a further shore / is reachable from here."

References

Amodio, D. M. 2014. "The Neuroscience of Prejudice and Stereotyping." *Nature* 15: 670–82. https://doi.org/10.1038/nrn3800.

Arain, M., M. Haque, L. Johal, P. Mathur, W. Nel, A. Rais, R. Sandhu, and S. Sharma. 2013. "Maturation of the Adolescent Brain." *Neuropsychiatric Disease and Treatment* 9: 449–61. doi: 10.2147/NDT.S39776.

Banaji, M., and A. Greenwald. 2013. *Blind Spot.* Delecourt Press.

Berezow, A. B. 2011. "Gender Discrimination in Science Is a Myth." *National Review*, January 12. https://www.nationalreview.com/2011/01/gender-discrimination-science-myth-alex-b-berezow/.

Bertrand, M., and S. Mullainathan. 2003. "Are Emily and Greg More Employable than Lakisha and Jamal? A Field Experiment on Labor Market Discrimination." NBER Working Paper No. 9873. National Bureau of Economic Research. http://www.nber.org/papers/w9873.

Campitelli, G., and F. Gobet. 2010. "Herbert Simon's Decision-Making Approach: Investigation of Cognitive Processes in Experts." *Review of General Psychology* 14 (4): 354–64. https://doi.org/10.1037/a0021256.

Carlton, G. 2023. "A History of Affirmative Action." *BestColleges.com*, July 20. https://www.bestcolleges.com/news/analysis/2020/08/10/history-affirmative-action-college/.

Ceci, S. J., and W. M. Williams. 2011. "Understanding Current Causes of Women's Underrepresentation in Science." *Proceedings of the National Academy of Science* 108 (8): 3157–62. https://doi.org/10.1073/pnas.1014871108.

Ceci, S. J., W. M. Williams, R. A. Sumner, and W. C. DeFraine. 2011. "Do Subtle Cues about Belongingness Constrain Women's Career Choices?" *Psychological Inquiry* 22 (4): 255–8. doi: 10.1080/1047840X.2011.619112.

Contreras, J. M., M. R. Banaji, and J. P. Mitchell. 2013. "Patterns in Fusiform Face Area Differentiate Faces by Sex and Race." *PLoS ONE* 8 (7): e69684. https://journals.plos.org/plosone/article?id=10.1371/journal.pone.0069684.

Cozolino, L., and S. Sprokay. 2006. "Neuroscience and Adult Learning." *New Directions for Adult and Continuing Education* 2006 (110): 11–19. https://doi.org/10.1002/ace.214.

Dasgupta, N. 2025. *Change the Wallpaper: Transforming Cultural Patterns to Build More Just Communities*. Yale University Press.

Devine, P. G., and T. L. Ash. 2022. "Diversity Training Goals, Limitations, and Promise: A Review of the Multidisciplinary Literature." *Annual Reviews of Psychology* 73: 403–9. https://doi.org/10.1146/annurev-psych-060221-122215.

Dickey Zakaib, G. 2011. "Science Gender Gap Probed." *Nature* 470 (7333): 153. doi: 10.1038/470153a.

Dixon-Fyle, S., K. Dolan, V. Hunt, S., and Prince. 2020. *Diversity Wins: How Inclusion Matters*. McKinsey Global Institute. https://www.mckinsey.com/featured-insights/diversity-and-inclusion/diversity-wins-how-inclusion-matters.

Donovan, T., K. Dunn, A. Penman, R. J. Young, and V. M. Reid. 2020. "Fetal Eye Movement in Response to a Visual Stimulus." *Brain and Behavior* 10 (8): e01676. doi: 10.1002/brb3.1676.

Duke University. Faculty Development. https://facultyadvancement.duke.edu/faculty-development/. Accessed September 11, 2023.

Eberhardt, J. L. 2005. "Imaging Race." *American Psychologist* 60 (2): 181–90. https://doi.org/10.1037/0003-066X.60.2.181.

Eberhardt, J. L. 2020. *Biased: Uncovering the Hidden Prejudice That Shapes What We See, Think, and Do*. Penguin Books.

Giedd, J. N., J. Blumenthal, N. O. Jeffries, F. X. Castellanos, H. Liu, A. Zijdenbos, T. Paus, A. C. Evans, and J. L. Rapoport. 1999. "Brain Development during Childhood and Adolescence: A Longitudinal MRI Study." *Nature Neuroscience* 2 (10): 861–3. doi: 10.1038/13158.

Golby, A. J., J. D. E. Gabrieli, J. Y. Chiao, and J. L. Eberhardt. 2001. "Differential Fusiform Responses to Same- and Other-Race Faces." *Nature Neuroscience* 4 (8): 845–50. doi: 10.1038/90565.

Goldstein, B. E. 2015. *Cognitive Psychology: Connecting Mind, Research and Everyday Experience*, 4th Ed. Cengage Learning.

Greenwald, A. G., and M. R. Banaji. 2017. "The Implicit Revolution: Reconceiving the Relation between Conscious and Unconscious." *American Psychologist* 72 (9): 861–71. http://dx.doi.org/10.1037/amp0000238.

Gretzinger, E., M. Hicks, C. Dutton, and J. Smith. 2024. "Tracking Higher Ed's Dismantling of DEI." *The Chronicle of Higher Education*. https://www.chronicle.com/article/tracking-higher-eds-dismantling-of-dei. Accessed September 18, 2023.

Gumbel, A. 2020. *Won't Lose This Dream: How an Upstart Urban University Rewrote the Rules of a Broken System*. The New Press.

Ito, T. A., and G. R. Urland. 2003. "Race and Gender on the Brain: Electrocortical Measures of Attention to the Race and Gender of Multiply Categorizable Individuals." *Journal of Personality and Social Psychology* 85 (4): 616–26. https://doi.org/10.1037/0022-3514.85.4.616.

Jones, D. S., D. D. Gillette, P. E. Cooper, R. Y. Salinas, J. L. Hill, S. J. Black, D. J. Lew, D. A. Canelas. 2022. "Cultivating PhD Aspirations during College." *CBE—Life Science Education* 21 (2): 1–19. doi: 10.1187/cbe.20-06-0111.

Kahneman, D. 2013. *Thinking, Fast and Slow*. Farrar, Strauss and Giroux.

Kandel, E. 1991. "Cellular Mechanisms of Learning and the Biological Basis of Individuality." In *Principles of Neural Science*, 3rd Ed., edited by E. R. Kandel, J. H. Schwartz, and T. M. Jessell, 1009–31. Elsevier.

Klawans, H. 2000. *Defending the Cavewoman and Other Tales of Evolutionary Neurology*. W. W. Norton.

Knox, L. 2024. "An Early Look at Diversity Post-affirmative Action." *Inside Higher Ed*, September 6. https://www.insidehighered.com/news/admissions/traditional-age/2024/09/06/early-look-racial-diversity-post-affirmative-action#.

Kupfermann, I. 1991. "Learning and Memory." In Principles *of Neural Science*, 3rd Ed., edited by E. R. Kandel, J. H. Schwartz, and T. M. Jessell, 1227–45. Elsevier.

Landivar, L. C. 2013. *Disparities in STEM Employment by Sex, Race, and Hispanic Origin. American Community Survey Reports, ACS-24. US Census Bureau.*

Li, D., and C. Koedel. 2017. "Representation and Salary Gaps by Race-Ethnicity and Gender at Selective Public Universities." *Educational Researcher* 46 (7): 343–54. https://doi.org/10.3102/0013189X17726535.

Lu, A. 2024. "Does Diversity Training Work?" *The Chronicle of Higher Education*, September 9. https://www.chronicle.com/article/does-diversity-training-work.

Mandela, N. 1995. *Long Walk to Freedom: The Autobiography of Nelson Mandela*. Little, Brown.

Masters-Waage, T., C. Spitzmueller, E. Edema-Sillo, A. St. Aubin, M. Penn-Marshall, E. Henderson, P. Lindner, C. Werner, T. Rizzuto, and J. Madera. 2024. "Underrepresented Minority Faculty in the USA Face a Double Standard in

Promotion and Tenure Decisions." *Nature Human Behavior* 8: 2107–18. https://doi.org/10.1038/s41562-024-01977-7.

Moss-Racusin, C. A., J. Dovidio, V. L. Brescoll, M. J. Graham, and J. Handelsman. 2012. "Science Faculty's Subtle Gender Biases Favor Male Students." *Proceedings of the National Academy of Sciences* 109 (41): 16474–9. https://doi.org/10.1073/pnas.1211286109.

Nolte, J. 1988. *The Human Brain: An Introduction to Functional Anatomy*, 2nd Ed. Mosby-Yearbook.

NSF (National Science Foundation). 2023. NSF's NCSES Releases Report on Diversity Trends in STEM Workforce and Education. https://new.nsf.gov/news/diversity-and-stem-2023#. 2023.

Padilla, N., and H. Lagercrantz. 2020. "Making of the Mind." *Acta Paediatrica* 109 (5): 883–92.

Pager, D., and H. Shepherd. 2008. "The Sociology of Discrimination: Racial Discrimination in Employment, Housing, Credit, and Consumer Markets." *Annual Review of Sociology* 34: 181–209. https://doi.org/10.1146/annurev.soc.33.040406.131740.

Ramuthi, D. 2024. "21 Eye-Opening Statistics about the Human Brain." *Venngage*, January 16. https://venngage.com/blog/human-brain-statistics/.

Reid, V. M., K. Dunn, R. J. Young, J. Amu, T. Donovan, and N. Reissland. 2017. "The Human Fetus Preferentially Engages with Face-Like Visual Stimuli." *Current Biology* 27 (12): 1825–8. https://doi.org/10.1016/j.cub.2017.05.044.

Rosenberg, K. R. 2021. "The Evolution of Human Infancy: Why It Helps to Be Helpless." *Annual Review of Anthropology* 50: 423–40. https://doi.org/10.1146/annurev-anthro-111819-105454.

Smith-Doerr, L., S. Alegria, and T. Sacco. 2017. "How Diversity Matters in the US Science and Engineering Workforce: A Critical Review Considering Integration in Teams, Fields, and Organizational Contexts." *Engaging Science, Technology, and Society* 3: 139–53.

Tversky, A., and D. Kahneman. 1974. "Judgment under Uncertainty: Heuristics and Biases." *Science* 185 (4157): 1124–31.

USSC (United States Supreme Court). 2023. *Students for Fair Admission v. Harvard*, 600 U.S. 181. https://www.supremecourt.gov/opinions/22pdf/20-1199_hgdj.pdf.

Wu, T., A. J. Dufford, M. A. Mackie, L. J. Egan, and J. Fan. 2016. "The Capacity of Cognitive Control Estimated from a Perceptual Decision-Making Task." *Scientific Reports* 6: 34025. https://doi.org/10.1038/srep34025.

Index

www.ingramcontent.com/pod-product-compliance
Ingram Content Group UK Ltd.
Pitfield, Milton Keynes, MK11 3LW, UK
UKHW020220290926
15470UKWH00001B/299

9 781666 975833